SS HELMETS

Kelly Hicks

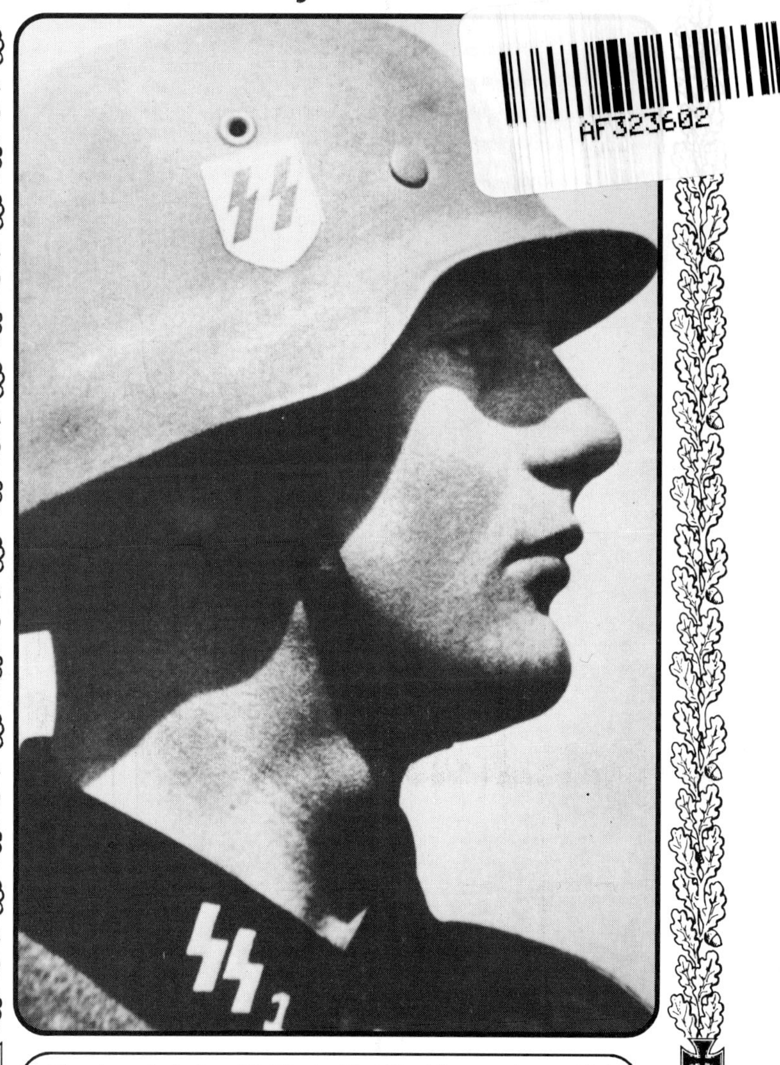

Classic portrait of an SS-"Deutschland" soldier wearing an M-40 steel helmet with 1st pattern runic shield. Note the "SS-1" collar tab still being worn.

Published by

Reddick Enterprises
P.O. Box 314
Denison, Texas 75020
(903)-463-1366
Fax (903)-463-7734

ISBN: 0-9624883-5-6

Edited by J. Rex Reddick
Designed by Ed Wells & J. Rex Reddick
Printed & Bound by Mix Printing Company

FRONT COVER PHOTOS
above: Allgemeine-SS M-35 satin black (originally field grey double decal)
below: M-42 single decal Waffen-SS with chicken wire for camouflage
BACK COVER PHOTOS
above: Earth brown Austrian M-16 transitional Allgemeine-SS
below: Satin black Allgemeine-SS Austrian M-16 transitional

TABLE of CONTENTS

ACKNOWLEDGMENTS

Many thanks to the individuals listed below, whose contributions and encouragement have made this book a worthwhile and enjoyable project. To my father, Edward Hicks, for sparking my interest, and a special thanks to my wife, Eun Hui, whose "tolerance" of the hobby has continued, steadily.

Al Barrows

Terry Goodapple

Russ Hamilton

John R. Angolia

George Petersen

R.J. Bender

Don Ginsburg

COL Ken Bowra

Ed Hicks

Ed Sayre

Jim Gauthier

Oswald K. Fernschild

Rex Reddick

Munin-Verlag

Mark Wills

I

INTRODUCTION

Collecting militaria has become a widely popular hobby throughout the world today. The thrill of owning a piece of history gives the hobby a special appeal which sets it apart from most other leisure time pursuits. To be sure, the values of some militaria have made the hobby into more than just a leisure time pursuit. For many, it is a total livelihood.

German helmets of both wars are an example of militaria which has risen enormously in value just over the last decade. Of the two, WWII helmets tend to be in greatest demand. Perhaps it is the variety of WWII helmets which spawns this demand.

With the help of some excellent works on the subject, all types of Nazi-era helmets, from paratrooper to RLB, have been documented in some detail. One problem with all this is that books which try to cover every type of helmet tend to leave information out. This guidebook is offered as a step forward in a bit more detailed look at helmets worn by the Schutzstaffel, or SS.

The subject of the SS attracts military history buffs in that it was not only one of the best fighting forces in modern history, but it also incorporated foreigners for more than half of its 900,000 soldiers. The 38 divisions of the Waffen-SS fought with distinction on all fronts of WWII, setting the stage for the employment of multi-national forces in modern warfare. In addition to combat units, this book covers helmets worn by the other non-combatant organizations of the SS.

The reader may assume that this work is purely historical in nature, and in no way condones or glorifies the activities of the SS in pursuit of Hitler's final solution.

The SS as a whole wore functional yet attractive uniforms. The helmets, with their distinctive runic insignia, are highly prized by many collectors. Original SS items are scarce, making them among the highest-valued German WWII militaria. One would think that with nearly 1 million men in SS uniform by 1945, SS helmets and uniforms would be more plentiful. For starters, they lost vast amounts of men and materiel in Russia and other parts of formerly denied areas of Europe. In addition, the order was given in late 1943 to cease issuing helmets with decals on them, as well as to remove all decals from those already in use. Although this order was obviously not followed to the letter, it surely helped reduce the number of helmets available to the collecting community which could be readily identified as "SS." Lastly, the desirability of SS memorabilia has caused most of it to be "collected-up" over the years.

The scarcity of SS helmets has caused a "skyrocketing" effect in their value. This, in turn, has led to the un-abated counterfeiting of them. The people who do the counterfeiting prey on all collectors, whether novice or advanced, stinging them for hundreds of dollars. Dealing with the counterfeiters is one of the key purposes for writing this book. To beat them at their own game requires teaching the collector to spot "real" helmets. There will be no "expose" on fakes found within this writing. Instead, there are only photos of real examples.

Sepp Dietrich wears the 1st. style LAH helmet. While not clear, the national colors shield would appear hand-painted. Note straight-line scallop. Al Barrows archives.

The reason for this is mainly to avoid confusion--the collector must focus on what a real helmet looks like, and the fakes will be easier to spot.

Therefore, this author's approach will focus on the characteristics of real helmets, devoting some discussion on comparing and contrasting real with counterfeit decals and finishes. The approach of this book is also to combine quality photographs of SS helmets with a liberal dose of common sense analysis.

Another view of the 1st style LAH helmet, showing the scalloped white on black runic shield. **Munin-Verlag**.

While not the definitive "last word" on SS helmets, this guidebook is a step forward in the clarification of some earlier information, plus an introduction of some "new" material. This "new" material consists of some already known, but not widely available information; and also some which has never before been documented. To ensure the quality and credibility of the material presented in this book, the author has consulted some of the noted experts in the helmet collecting field.

This has proven of tremendous value in establishing sound analysis where the photographic or documentary record from the WWII period is lacking. Wherever possible, photographic support for the material presented in this book is used. Hopefully, collectors at all levels can find a use for this guidebook toward identifying helmets in their own collections or ones they run across in their searches. Finally, it is hoped this book will help enhance the collector's overall level of knowledge, enabling him to collect SS helmets with confidence.

Men of the LAH wear the national colors shield on the left side of their helmets.

II

"STANDARD" ISSUE SS HELMETS

A BRIEF OVERVIEW

1933-1935; the Allgemeine-SS

Upon Hitler's assumption of power in 1933, the SS began to grow in numbers beyond just a bodyguard formation to a complex political institution. Although somewhat small compared to the SA, the SS was becoming diverse nonetheless. Its organizations consisted of the Leibstandarte SS Adolf Hitler, the SS-Totenkopfver-bande, the SS main offices, and the district units of the Allgemeine-SS.

A formation of the Leibstandarte on parade, 9 November, 1935, in Munchen. They are wearing the standard 1st. pattern runic shield. **Hicks photo collection.**

Eventually, as a paramilitary role was identified for the SS, recruitment was begun for the new SS-Verfugungstruppe. Helmet requirements, primarily for parade functions initially, now began to include the demand for field purposes as well.

The only helmets available at this time were surplus WWI helmets. With Germany in the midst of a great economic depression, these surplus stocks were put to their fullest use by the entire armed forces. Since the helmets were relatively new (15 to 17 years old), many did not require refurbishment in the form of new liners and finishes prior to being issued directly to units. As a fledgling, para military organization the SS had little status next to the Army or Navy when it came to helmet procurement. The supply systems of the traditional armed forces were much more complex institutions, which enjoyed the support of the the German defense establishment at the highest levels. The SS, on the other hand, was regarded as a band of outsiders who had no real military role. As a result, they had to accept whatever they got from the supply system.

By 1934 the SS was incorporating civic and "medium duty" style helmets to augment their limited supplies of leftover WWI helmets. Additionally, the Reichszeugmeisterei (RZM) produced a special SS helmet in 1934. These three types of helmets were the only ones which were produced in black for the Allgemeine-SS (dark blue-grey for RZM helmets). All others which were supplied to the SS arrived with a field grey finish, and had to be repainted black to match the SS uniforms. This accounts for most Allgemeine-SS helmets displaying multiple layers of paint. Sometimes, the inner domes of these helmets were left field grey.

Men of an unidentified SS-TV unit stand inspection. They wear helmets with stenciled or hand painted death's heads. **Bender**.

Insignia used by the SS during this early period were non-standard in both design and manufacture. The Leibstandarte utilized a scalloped shield with silver runes on a black background. A tri-color national shield, both scalloped or unscalloped, was affixed to the left side.

Within a year, the Leibstandarte went to Black runes on a silver unscalloped shield, maintaining the tri-color shield on the left. The SS-TV (Death's Head units) some-times used a stenciled skull and crossbones emblem on their helmets. The SS-VT (Special Purpose Troops) wore a circular emblem with runes on the right, and a mobile swastika on the left. RZM SS helmets were initially issued and worn without insignia of any kind, and their use (without insignia) appears to have been fairly wide throughout the various SS formations for about a year. According to the best sources of both documentation and photographs, these earliest styles of SS insignia could either be hand-painted or decals.

Because so few originals remain, the collector must be absolutely certain of the helmet's "lineage" before invest-ing his money. The reason so few exist is one of supply; as insignia were standardized for the entire SS, the non-standard helmets were re-fitted with new paint and insignia as quickly as possible to maintain uniformity.

In Mollo's Uniforms of the SS, vol. 3 is another clue to the fate of not only the earliest helmets, but of many SS transitionals in general. He refers to an order given by the Economic and Administrative Main Office in March, 1941 which directed the old model (transitional) helmets be sent to the penitentiary at Straubing, Bavaria for melting-down.

1935-37; Period of Standardization

In August, 1935, SS Helmet insignia were standardized in the form of the widely-recognized runic and swastika shield combination. A popular producer of the decals was the firm of C. A. Pocher of Nurnberg. The runic shield, commonly referred to as the 1st pattern shield was produced by printing black runes and a border against a shield of finely-granulated aluminum. The swastika shield consisted of several variations, with the basic characteristic of the black swastika inside a white circle against a black-bordered orangish-red shield. Variations exist in the "fatness" or "thinness" of the swastika (see section VII). Also, on most original party shields, the white circle is slightly right of center (this is quite visible upon close examination).

These high quality decals were printed on very thin celluloid material, which gave them the ability to "snuggle" into the surface of the helmet. By the end of 1935, they had become the standard helmet insignia for the entire SS. The fate of the thousands of helmets with the non-standard insignia of the earlier years was thus sealed. Units such as the Leibstandarte, who had undergone two previous insignia changes, were again re-finishing their helmets and applying new insignia, thus obliterating all traces of the earlier pattern SS shields. This standardization took place quickly and comprehensively, overlooking conceivably very few helmets in the process. Thus, helmets with the old styles of insignia should be regarded as extremely rare.

During early 1936, a new style of steel helmet, called the M-35, began making its appearance in SS units. The

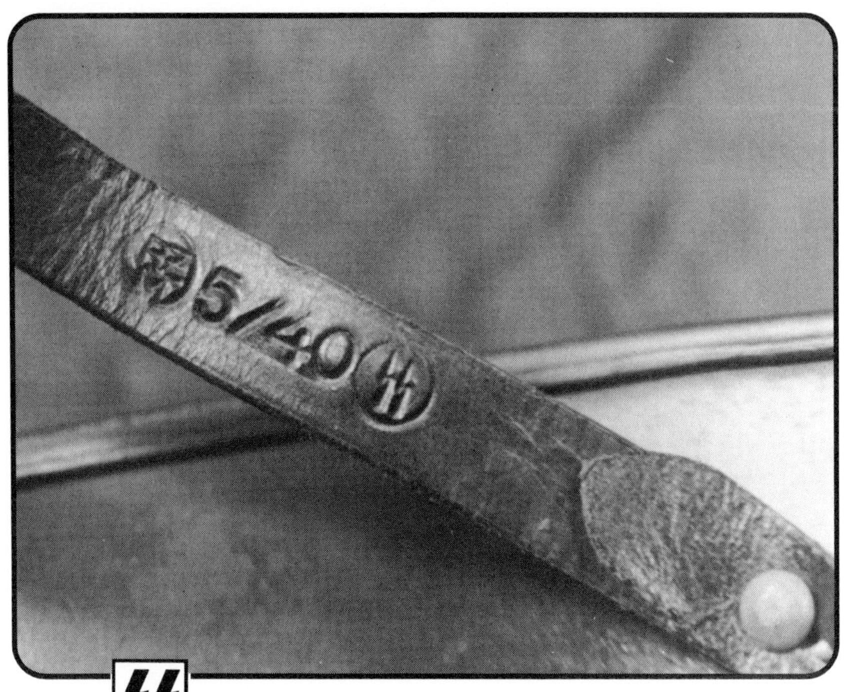

Detail of RZM marking on SS chinstrap. **Hicks collection.**

M-35s originated from various makers, and were supplied to the SS through the standard Wehrmacht channels. It has been postulated that only one or two helmet manufacturers (such as E.T. or Q) produced solely for the SS. This theory is real tough to support. Within a military-industrial complex as huge as Germany's was during the re-armament period, manufacturers supplied centralized Wehrmacht procurement offices based on requisitions. It is therefore inconceivable that there was any deliberate targeting of Army, Navy, or SS by specific makers; rather, that helmets were distributed from depot to unit on an "as needed" basis. Moreover, all makers are represented in original M-35 SS examples (N.S., E.T., Q, S.E., and E.E.).

As the M-35s began arriving at units, they were re-painted (black) and given SS insignia. For those SS units with specific combat readiness roles, the helmets remained field grey. As the supply of M-35 helmets was limited initially, all types of helmets heretofore mentioned (M-16, M-1, M18, and RZM) saw concurrent use through 1938. Photographic evidence corroborates this, showing mixtures of M-35 and transitional helmets being worn side by side.

By 1937, a 2nd pattern SS runic decal had made its appearance. This design differs from the 1st pattern in that the runes are "blockier," the shield is slightly less metallic, and the border of the decal is wider. This pattern of decal appears only on field grey helmets, a fact which indicates it was not widely used until the late 1930s (note that helmets with dome dates as early as 1937 have displayed 2nd pattern runic shields).

1937-1940; The SS-VT and Waffen-SS

This period saw the most rapid expansion of the SS during the pre-war period. The impact this expansion had on helmets was again one of standardization (to the M-35 pattern for good). The SS as a growing military force was moving away from parade and guard activities and more toward field training. More and more M-35s were arriving through the supply system. Transitionals were fitted with the M-31 liners characteristic of the M-35 helmets.

Allgemeine-SS helmets were repainted field grey (and new decals applied) to conform with the earth-brown and later field grey uniforms of the SS. This is not to say black helmets fell from use. On the contrary, black M-35s appear in photographs dating into late 1939. At that

Waffen-SS recruit pictured wearing an M-40 double decal helmet. These helmets were issued with or permitted to display double decals until late 1940. **Bender**.

same time, however, the number of transitional models seems to be almost nil, except for reserve and "wachtbattalion" use.

Upon the annexation of Austria in 1938, Austrian SS units were standardized within the greater German SS structure. With this came the standardization of insignia for their helmets. While some Austrian units used German-produced insignia prior to the annexation, others appear to have used locally manufactured styles which approximate the German (C.A. Pocher) design. Some of these designs, which include variant party shields, can still be found on existing examples dating from the period.

The first SS-VT units to rush across the borders of Czechoslovakia, Poland, and France wore field grey M-35 helmets with double decals and parade quality finishes. Regiments such as "Germania" had not yet been issued camouflage clothing and resorted to using mud and vehicle paint (a practice which continued well into the war) to subdue their helmets. This was in direct response to the troops' need for concealment on the battlefield, and helmets began to be produced with rough texture or non-glare finishes.

This modification was incorporated into the overall design of the M-40 combat helmet, which also featured a stamped (instead of riveted) vent hole. Additionally, the chin strap "d-rings" in the M-31 liner were rounded at the corners. M-40 SS helmets exhibit either pattern of runic shield, still worn on the right side of the shell.Until the order was given in late 1940 to discontinue the use of the swastika shield on the steel helmet, the SS wore both decals on the rough finish M-40 helmet.

Commander of the SS Flakabteilung Obersalzberg, Ostuf. Emil Kurz, wears a "classic" M-35 double decal helmet with field grey parade finish. **J. R. Angolia**

1941-45; the Waffen-SS

In terms of standard issue helmets, the only change for the remainder of the war was the introduction of the M-42 steel helmet. The M-42 differed from the M-40 only in that the rim of the helmet was left uncrimped as a time-saving step. The liner was still the M-31. Until October, 1943, when the Reichsfuhrer-SS ordered the removal of all runic shields from helmets, these helmets continued to incorporate both styles of runic shield. Himmler's order, official as it may have been, seems to have been largely ignored, as so many original SS helmets still retain their decals.

III

VARIATION SS Helmets and Insignia

Presented in the previous chapter was a brief synopsis of the helmet and insignia types in standard use by the SS between 1933 and 1945. In this chapter, several of the most often encountered variations of SS helmets are presented. This information is completely supported in both photographic and artifactual evidence. Moreover, while some of the examples herein have been discussed in previous books on the subject, there has not to date been as complete a treatment of variation SS helmets.

"Reversed-Decal" Helmets

These are SS helmets which have an SS runic decal on the left side, and a swastika shield on the other. The predominant types of helmet displaying this decal placement are the M-34 Medium Duty or "SD" helmet, and the M-42 combat helmet. These helmets may be affiliated with police or security roles, based on photographic evidence showing them in use by uniformed German and foreign field police. The fact that many SD helmets exhibit reversed decals is no surprise considering the nature of their mission and police affiliation. Moreover, the SD, like the police, wore double decals throughout the war.

Some difficult phenomena to explain are the instances where SD helmets exhibit reversed decals with addi-

Not all foreign volunteers wore SS runes on the opposite side of their helmets! This "Langemarck" volunteer wears an M-35 with properly placed 1st. pattern shield. **J.R. Angolia**

tional decals on top of them in proper configuration. Possibly, an order was given for all SD members to wear helmet insignia in accordance with SS procedures. Although logical in terms of the "dual status" of the SD as SS or police, this is only conjecture based on existing original examples which display these traits.

M-42 helmets which fall into the reversed decal category appear for the most part to have been combat police helmets with SS runic shields placed directly over the police decal. There is photographic documentation of this type in use, but no known official order exists governing their purpose.

Italian members of the 29th SS Division receiving awards. Note the man in Schutzpolizei uniform with runic shield on the left side of his helmet. **Barrows**.

One explanation for these helmets is that when German and foreign Schutzmannschaften were standardized under the SS, it was easier to add the runic shield on top of the police emblem than to use two decals (runes and swastika). Another possibility is that the configuration of reversed decals was in keeping with the police nature of their organization.

"Double Runic" SS Helmets

Although extremely rare, enough double runic helmets have popped-up to spark this author's interest. In at least two photographs, members of the Leibstandarte are shown wearing helmets with runic shields on both sides. These helmets tend to be M-35 shells, with anything from a textured camouflage to a parade finish painted over the original factory finish. The standard runic shield is either masked-off, and a runic shield applied over the swastika shield, or the whole helmet is overpainted and two new runic shields applied on top of the finish. No official documentation explaining this variation exists, so it is difficult to pinpoint the real purpose for the double runes. Many experts feel that these were helmets worn by foreign volunteers, who either under orders or by preference chose to cover their swastika shield with a runic decal.

Generally speaking, however, by the time foreign units made their appearance in large numbers, the M-42 helmet, not the M-35, was standard issue. Since most double runic helmets tend to be M-35s, it is likely their use predated the inception of most foreign units. It could simply be vanity which caused soldiers to put another runic shield on their helmets, as some appear to have no party shield under the second runic decal. This is logical when it is remembered that in 1940 the order was given to eliminate the swastika shield from use. When units stood down to repaint and "single decal" their helmets, some individuals may have chosen to use the extra runic shield they were issued to place on the other side of the helmet.

Again, conditions during wartime are "foggy" and ever-changing. What came out as a general order cannot be guaranteed as having been fulfilled to the letter. Moreover, for esprit purposes, individual soldiers and whole units for that matter have been known to modify or add to their uniforms in a distinctive manner. Since decals were available down to the lowest levels, the odds are fairly high that soldiers on occasion took their own initiative with their helmets.

Men of the Leibstandarte in operations in Russia. Note the runic shield on the wrong side of the helmet on the soldier in the foreground. **Munin-verlag**.

"Mirror Runic" SS Decals

Existence of a photograph, as well as an original specimen, brings to light an odd variation SS decal, the "mirror rune." The close-up example shares similarities with the Austrian variation runic shield, i.e. a thicker border and a slightly smoother metallic background (as compared to the C.A. Pocher decal, which has a visibly granular metallic texture). The fact that the original mirror specimen shares characteristics with the "Austrian" variation points to its being of Austrian origin as well.

Flemish SS volunteer Remy Schrynen wears a mirror runic shield on the left side of his helmet. Whether these decals were "mistakes" or deliberately manufactured backwards is a mystery. *Barrows*.

However, it is unknown why the runes are printed backward. In some instances, Luftwaffe eagles appear flying backward on parade style helmets and are thought to be private manufacturers' variations. This is quite likely for the mirror rune as well. Another possibility is that the decal is a mistake, and was used anyway rather than being discarded.The fact that a mirror runic shield shows up in a picture of a Belgian volunteer in the "Langemarck" regiment (see illustration) is also curious, suggesting a possible foreign use. A possible connection with mirror runic collar tabs must also be considered, but again, documentation is lacking. Whatever the purpose, there could not have been very many mirror runic SS decals produced, and original examples must therefore be considered extremely rare.

"Foreign Volunteer" SS Helmets

The photographic record is full of examples of various foreign troops wearing the SS runic shield on the left side of the combat helmet. This practice probably was the result of official orders governing the wearing of SS insignia in general by those whom Himmler considered to be "undesirable" foreigners (vis a vis special non-runic collar patches for many units). Examples of foreign volunteer helmets could conceivably appear on any style combat shell, but most likely would be on the M-42 because of the relatively late period of induction of foreign volunteers on a large scale. There are likely some variations among foreign insignia, as well as paint schemes, etc. The "rules" that apply to judging originality in standard German helmets should apply equally to the foreign examples.

SS Paratroopers wear standard, non-insignia Luftwaffe paratrooper helmets. **Bender.**

"SS Paratrooper" Helmets

By 1943-44, when the SS paratroop units were being employed against Tito and in the Mussolini rescue, purely luftwaffe stocks of paratroop equipment were utilized. Although examples of SS camo pattern smocks exist, the same does not appear true for the helmets. The small size and unconventional nature of SS paratroop units and operations cannot have generated formal issue of SS paratrooper helmets per se, but it is possible, again based on individual preference, that a helmet or two had runic shields applied.

For the most part, the lightweight M-34 "medium duty" field grey helmet is visualized when "SD" is mentioned. In truth, several varieties of helmets can be attributed to that organization. Photographic evidence shows lightweight civic, lightweight transitional, and combat shells being worn. What sets them apart from "run of the mill" SS helmets is the tendency for the runic and swastika shields to be "reversed." Many examples show runic shields placed directly over swastika shields, and vice-versa. This leads some experts to conclude that the SD identified more as a police organization early on, hence the swastika shield being worn on the same side (right) as on police helmets. As time progressed, the SD followed the same decaling procedure as the Waffen-SS for uniformity.

IV

Judging Originality of SS Helmets

*K*nowing *what to look for* is crucial to the collector and forms the key intent for this book. There are so many counterfeit SS helmets as to boggle the mind. This is unfortunate, but there is still a way to differentiate the fakes, no matter how convincing they appear. Even helmets with newly applied original decals can be detected if the collector understands the basic characteristics of both original and altered samples. This chapter will focus on four categories of condition: fake

Familiar "Signal" magazine photo of a Dutch SS volunteer wearing a 1st pattern shield on the left side of his helmet. **Al Barrows Archives**.

decals, fake finishes, newly applied original decals, and original, period-applied decals.

Fake Decals

As mentioned in Goodapple, Vol. 2, some reproduction decals are extremely well made, and except for size, almost cannot be distinguished from originals. Nowadays, proportionally correct versions are available, for a few dollars, which can be very convincing, especially when combined with a good "camo" job. The one key element missing from even the most technically "correct" decals is the metallic content. Original decals, which will be discussed more fully later, possess a beautiful metallic luster which (so far) is simply iimpossible to duplicate. If in doubt about what constitutes "metallic content," a trip outside into the sunlight will reveal the original decal's dazzling characteristics.

Even the more subdued-looking 2nd pattern runic shields will show a remarkable amount of glitter. Repro decals not only fall far short of this quality, they also are typically made of plastic, which makes them easy to spot for a number of reasons. First, the silver SS shield is darker grey than the originals. Second, plastic shields curl when damaged, a characteristic not shared by celluloid originals (which flake when damaged). Third, since plastic decals incorporate enamel or paint, they are thicker than originals, and require a lot of grinding before they will "snuggle" into the paint like an original.

Fake Finishes

One of the most common fake finishes is the camouflage finish. Every time an "artist" makes one of these, he is committing the disgusting act of ruining the originality and historical value of a once-original military artifact. Not only that, he is creating one of the most effective forms of trickery among all malicious faking techniques. This is due mainly to the fact that so many different types and colors of paint can be used. Mixed with dirt, sand, and other materials--then "aged" or abused, these paint schemes can look quite convincing. The easiest approach to detection, as Goodapple mentions, is to smell the paint. An odor of paint will immediately tell the collector the finish is not original. A finish which shows no abuse is also suspect to some degree in that helmets were camouflaged ostensibly to be used in combat. Combat use is very rough on the finish and liner of a helmet, and the effects of exposure to the elements causes unmistakable wear on both.

As a general rule, look for a camouflaged helmet to show rust and flaking to the finish, with the decals at least partially covered up, as well as commensurate dryness and darkness to the liner.

Other finishes can be altered or added to fool a buyer. The most commonly encountered of these is the "picnic table" stain finish, designed to "age" a repro decal or camo finish. This staining is particularly damaging to the helmet's original finish, and is ironically one of the easiest alterations to detect.

Common sense must guide the collector in evaluating a helmet. Smelling the surface of the paint is an excellent habit to get in to despite its being an unusual approach. However, a 20-year-old fake may have no odor, so other discriminators must be looked for, such as decal characteristics. It is also important to keep in mind the Germans used whatever was on hand at the time. Therefore, paint quality may vary (be of very poor quality, for example) but still be "good."

Period photos of an Einsatzkommando raid in Amsterdam. Note the model M-34 medium duty helmets in use. **B. Smith.**

Newly-applied Original Decals

The "state of the art" fake helmet is the no-decal M-42 converted to a single-decal Waffen-SS through the application of an original decal. There are many philosophies surrounding whether this is an "original" helmet or not, but the bottom line is that no SS trooper wore it. For those who collect for the historical importance of it, this makes a big difference, especially since the natural inclination of the "artist" is to demand the same price for the helmet regardless of the fact it is a "parts" helmet.

Discerning a newly-applied from a period-applied decal can present a big challenge to the collector. On a well-done job, there is only one main difference which is the brightness of the shield. This brightness is unmistakable, and measures taken to mask it are also usually obvious. These can include a combination of staining, abrading, and heating, all of which are detectable.

Decals which have blown apart during application (most do because of age) are a good indicator of recent application and will display a "pieced-together" appearance, with lines running across in different directions (It must be noted that many original period decals will show some tearing or wrinkling--generally not as much as recently-applied ones). In Ludwig Baer's book is a translated section of a period decal applying instruction booklet, which discusses every type of problem that might be encountered while applying decals. The translation clearly indicates that cracking, wrinkling, or bubbling also occurred in the old days, and should alert the collector not to pass on every helmet which has flawed decals. Again, the "common sense" test must play a major role in judging a helmet!

Period-applied Decals

Up to this point, the main points of repro decals and finishes have been discussed. The question, "How does one recognize the original?" still remains. This is not an easy task to accomplish. Generally, there is no substitute for years of experience in handling original helmets. The range of helmet conditions alone calls for vast amounts of information which cannot be contained in a book. Much like learning to drive a car--at some point, there has to be some "hands-on" training to augment what is in the manual. The best anyone can do is offer guidance about what to look for, and hope those readers who have not had a lot of experience handling helmets will be able to apply that guidance successfully.

As with every other process, there are rules which apply to helmet evaluation as well. Those listed below are important, but by no means hard and fast--they are somewhat "flexible" rules. Their purpose is to help the reader foster an analytical frame of mind in order to judge helmets more objectively.

(1) Regulations serve as guidelines for conduct within a military organization. Units and individuals often do not follow them to the letter. This is especially true under the constantly changing conditions of combat.

(2) The SS as an institution encompassed a huge variety of civil, political, and military organizations, both German and foreign. These organizations wore an immense array of uniforms and insignia.

(3) Decals were widely available and were applied on helmets at every level from depot to individual user. As such, they can exhibit a wide range of skill in application.

(4) There is no substitute for good, honest age on a decal or paint finish, whether used in combat or left in a barracks after the final parade.

(5) Getting overly excited at the prospect of owning a rare helmet can cloud judgment.

Keeping the "rules" in mind when inspecting a prospective addition to the collection could be helpful avoiding misjudgments. The discussion in the remainder of this chapter will focus on decal characteristics for each type of SS helmet.

Allgemeine-SS Helmets, pre-August, 1935

The LAH

Leibstandarte helmets of all WWI patterns will exhibit a layer of satin black, either sprayed or brushed, over the WWI or mid-30s field grey finish. Helmets displaying satin black over "Austrian brown" also existed. Upon examination of an original Allgemeine-SS transitional helmet, the collector will notice that almost invariably, the original 3-pad WWI liner is still in the helmet. Often, the black overspray will have gotten onto the edges of the liner and steel (or leather) band. Also, the inside dome of the helmet may remain field grey. The liner may be a mid-30s produced copy of the WWI 3-pad variety, but the construction is generally not as sturdy as the combat version. Chin straps can be the original two-buckle WWI strap, or 30s vintage copies, in either brown or black. Some have the heavy duty buckle with roller-bar.

The special LAH insignia (scalloped runic shield and national colors) may be decals or hand-painted. If the latter, then age cracking and yellowing should be visible and quite uniform throughout the emblem. Unevenness of color may occur, but should never be too extreme, as this (along with marked paint bubbling) is a characteristic of heat treatment to simulate aging.

SS troopers in France with mud-covered helmets for field expedient camouflage purposes. **Munin-verlag.**

SS-Verfugungstruppe

Deutschland, Germania, Der Fuhrer,

Signal, Artillery, Engineers, etc.

Between the LAH and the SS-VT there is no difference in the basic helmet characteristics. SS-VT helmets display the runes within a circle on the right, with a white-outlined mobile swastika on the left. Variations to this design exist (Goodapple, Vol. 2).

Most of these insignia were probably hand painted or stenciled, although limited numbers of decals may have been produced in these designs. Whether decals or painted, these insignia should possess a moderate to high degree of quality.

SS-Totenkopfverbande

Photographs dating from the pre-August 1935 period show SS-TV troops wearing the large death's head stenciled on the left side of their black helmets.

Allgemeine-SS, SS-VT and SS-TV

August, 1935 - October, 1939

Black or "Allgemeine" transitionals, as mentioned above, displayed variable quality satin black finishes on top of field grey. During this period, it is possible for a helmet to have had a 3-pad or M-31 style liner. Decals are the standard 1st pattern runic shield and swastika shield combination. On these parade helmets, the decals generally show little wear. The runic shield, if lacquered, will show a light "brass" coloration. Unlacquered examples can exhibit similar age toning, but will mostly retain a nice whitish-silver color.

Additionally, they will tend to show more flaking than the lacquered shields. Close examination will reveal the rich metallic content of the shield. In most cases, the edges of the decal can be felt, as they are lying atop a smooth finish, and do not "snuggle" as much as on a rough texture finish. While these helmets were mainly parade and guard helmets, and therefore should not show the abuse of their combat counterparts, they should exhibit some other clues that they are almost 60 years old. The WWI liners adapted for use in the 30s can be particularly age-worn or damaged. Many show contemporary repairs or reinforcements to the leather. Helmets which had a lot of use can show much flaking to the top, from being "put-down" a lot. Some of the poorer quality repaints can be badly flaked, revealing the underlying field grey paint finish. Bare metal which is exposed should be browned from age, or perhaps slightly rusted (Fresh rust could be a sign of tampering intended to age the helmet. This rust is more orange in color, and will rub off easily to the touch).

Many, if not most, black helmets were eventually re-entered into the supply system and re-finished for use as field helmets. Those which survived the early re-uses and were not turned in for melting down at Straubing or other locations often became subject to use by volksturm and luftschutz units late in the war.

The field grey transitional and M-35 helmets used by the SS-VT, TV and LAH were either repaints, as mentioned, or unaltered field grey combat shells drawn directly from the supply system. For the most part, the insignia were the standard 1st pattern runes and swastika shield. Many M-35 field grey helmets with dome dates of 1937 and later display the 2nd pattern runes. So far, this author has been unable to document a black or field grey transitional SS helmet with 2nd pattern runes. If they existed (logically, why shouldn't they have?), they have proven elusive in both photographic and material evidence.

The practice of lacquering helmet decals is much more visible on the field grey shells. The Germans used several types of lacquer, one of which went by the brand name "Ducolux." The purpose of applying the lacquer was to enhance adhesion of the decals, as well as protect them from exposure to the elements. On mint examples, the lacquer appears as a semi-glossy coating, either brushed or sprayed-on. It is almost always yellowed with age. This is a major cause of the gold tone which SS decals assume over time. In most cases, the lacquer will overlap the border of the decal, causing a very subtle change in the color of the helmet finish it covers. This appears as a light outer border around the decal.

On all lacquered examples, regardless of the degree of wear to the helmet, this lighter edge can be discerned, even after all the lacquer appears to be worn off from exposure. In most cases, the slightly brown-colored glue used to apply the decals is also visible just within the borders of the top coat of lacquer. While these characteristics are fairly universal on helmets, their absence does not necessarily indicate a fake. Again, some helmets were not decorated precisely according to the "rules." A final point about decal lacquer; many times the party shield will exhibit "spider-web" cracking, while the runic shield will not show any.

This is probably because the non-metallic composition of the party shield, vs that of the runic shield, allows for greater susceptibility to the effects of the lacquer. In many cases, unlacquered decals seem to fare better after all than those with the protective coating.

Waffen-SS, 1940-45

The M-35, M-40 and M-42 helmets in use during this period continued to display both patterns of runic shield. During 1940-41, the SS-Polizei-Division wore a mixture of double decal police M-35 helmets (unbordered police shield) and standard SS helmets. The only method of identifying whether or not a field police helmet is an "SS police helmet" is if it is so marked. By early 1942, the division's replacement personnel and equipment were coming directly from the overall SS supply system and their lineage as a police formation was mainly obliterated (just as the case with "Totenkopf," they were police in name only thereafter).

Aside from the SS-Polizei-Division, which looked like any other Waffen-SS formation after 1942, there were police formations which later became standardized into the SS from various schutzmannschaften and feldpolizei origins. These units performed rear area security duties such as einsatzkommando or feldgendarmerei. They often wore a mixture of police and SS insignia, including on their helmets. Many examples exist of double decal police helmets with an SS runic shield placed directly over the police shield.

Others appear with SS insignia (both decals) placed atop the police insignia in proper arrangement. Many collectors balk at helmets with multiple or oddly-placed decals, but as long as they conform to the basic characteristics of age and originality, they should not be overlooked. Moreover, as one begins to investigate the history of the SS and police organizations during the war, he will find that the absorption of security forces into the SS command and control structure happened on a large scale. The war-time supply system being what it was, however, not all inductees had the luxury of outfitting themselves in SS kit from head to toe. More realistically, a runic decal or a sleeve eagle had to suffice to make the man "SS." In terms of authenticity, the collector must judge for himself whether or not the criteria of age, wear, and proportion (dimensions of the decal) are met when viewing a variant piece. Before passing a helmet by because a decal is applied over the top of the camouflage or re-paint surface, think carefully about the "rules" mentioned earlier in this section. Consider the helmet as a soldier would--functional, yet part of a uniform; to be kept looking sharp as well as just in good working order.

Regarding the "mainstream" SS helmets of the war period, there is little about them which requires in-depth analysis. Both patterns of SS decal were used extensively on both the M-40 and the M-42. There are, however, some general guidelines which can be attributed to these helmets:

(1) They show much more wear than parade examples.

(2) On rough-texture finishes the decals adhered poorly compared to the smooth finishes. This occurred as a result of the person applying the decal not allowing the fixative to set-up before applying the decal, which led to slipping or non-adhesion in some parts of the shield. These areas later flaked or popped off quite easily. Another cause for lack of proper adhesion was the tendency

for dirt or air pockets to form within the rough texture of the finish.

(3) Because of the thinness of the decals, they quite naturally assumed the characteristics of the paint finish, literally becoming part of it. These are the so-called "decals which cannot be felt."

V

CONCLUSION

This book has been presented in hopes of providing collectors at all levels an additional reference tool for authenticating helmets worn by the SS and its associated organizations. By far not the "last word," this book is instead another piece of the puzzle, hopefully with more clues for enthusiasts to add to their knowledge base. It has been written with a high degree of confidence in the accuracy of the material presented, but, as with most other writings on the subject, this guidebook may contain information which some may dispute or simply have more knowledge on. In either case, criticisms are most welcome, as this adds to the learning process for everyone.

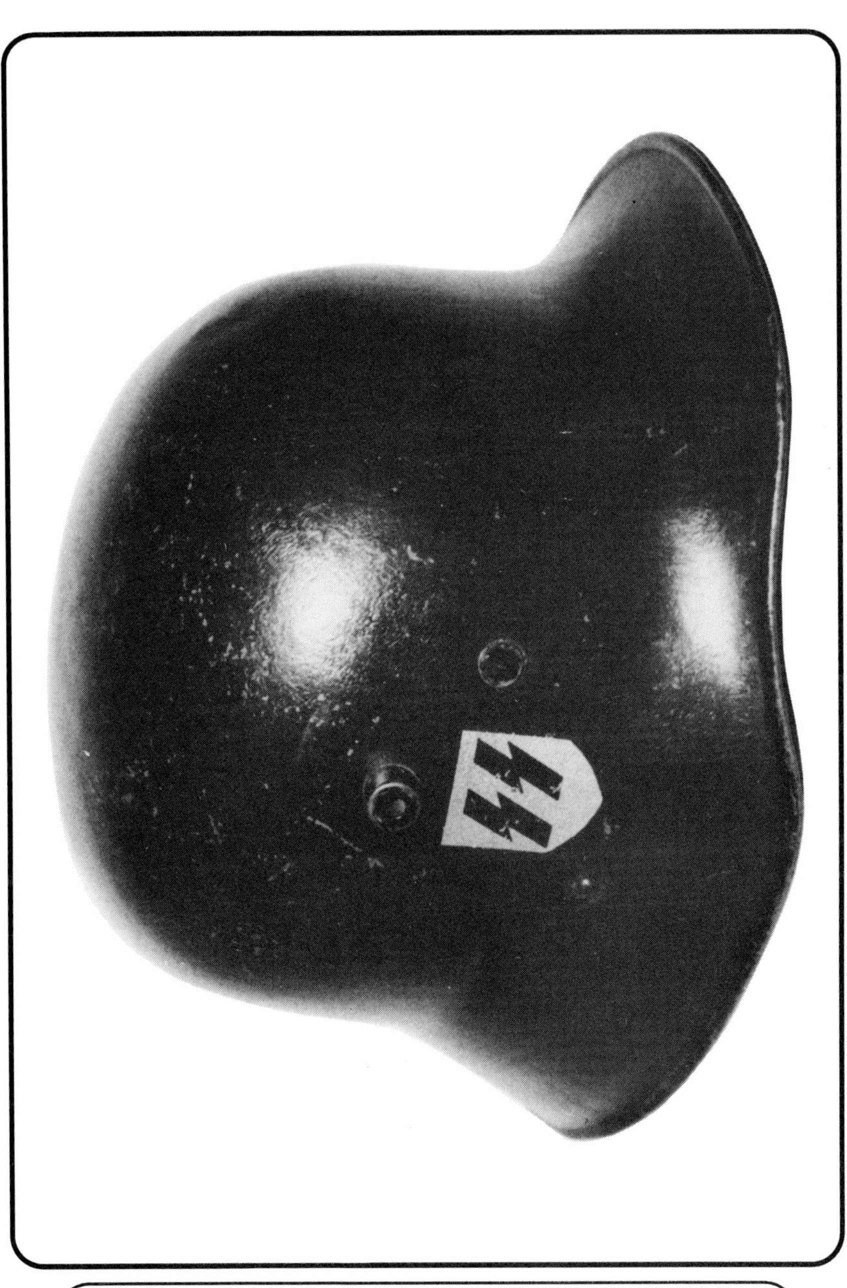

*Runic side view of an Allgemeine-SS helmet. The shell is an Austrian M-16, and the finish is satin black over the field grey. The decals are the standard C.A. Pocher make. **Hicks collection**.*

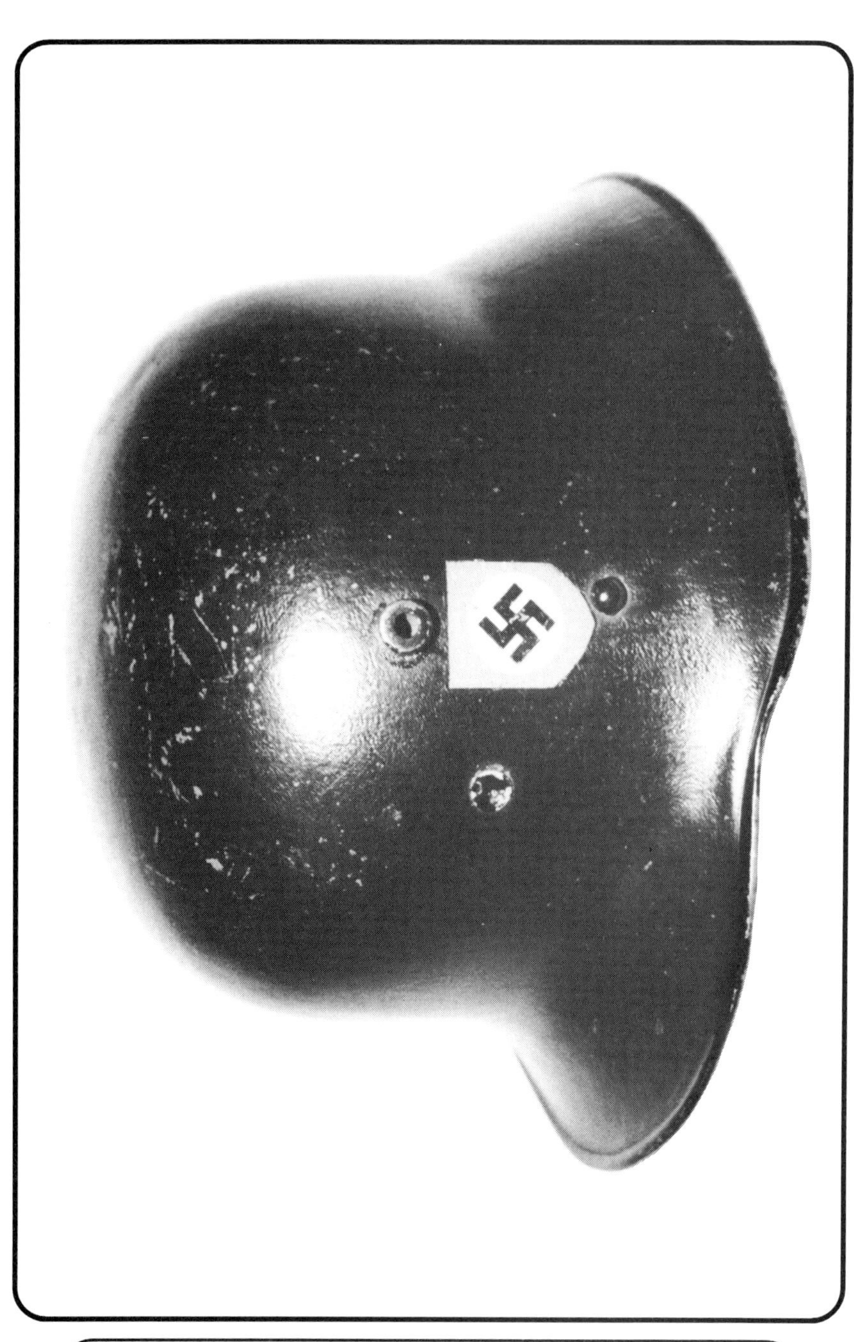

Party shield side of Allgemeine-SS helmet. **Hicks collection.**

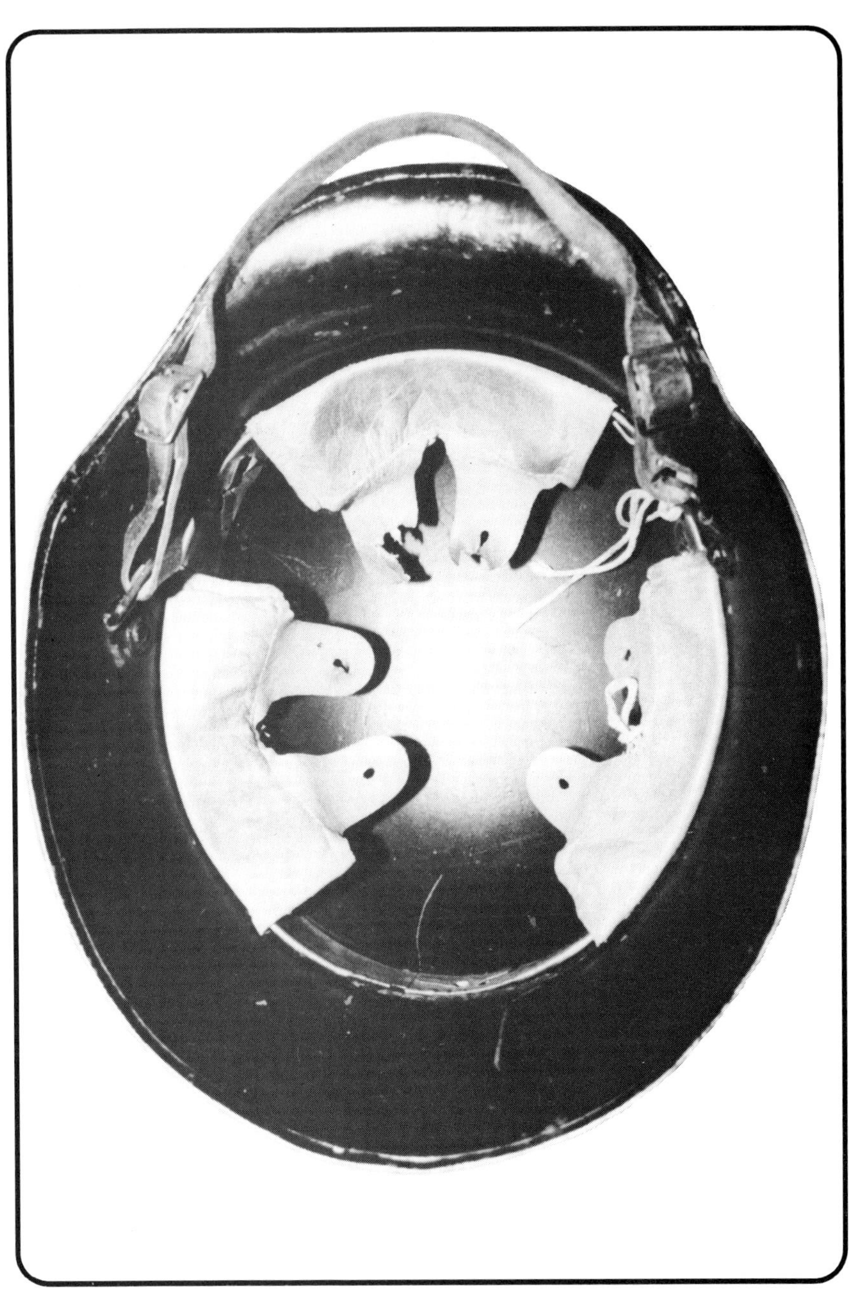

Interior of Allgemeine-SS helmet, showing satin black finish throughout. The liner is of mid-1930's manufacture. **Hicks collection**.

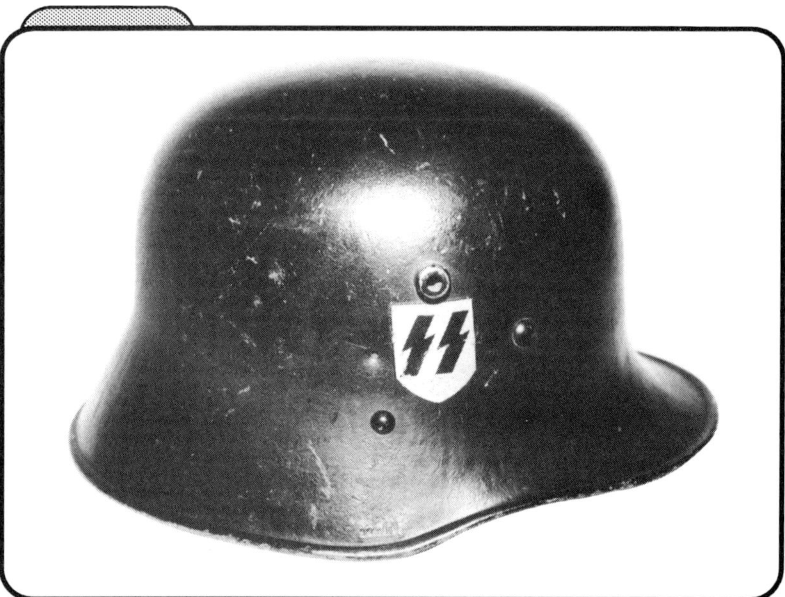

UPPER: *Allgemeine-SS helmet consisting of an Austrian M-16 shell with a satin black over field grey finish. The apertures for affixing WWI period insignia are visible near the 1st. pattern runic shield.*

LOWER: *Left side of same helmet showing party shield.*

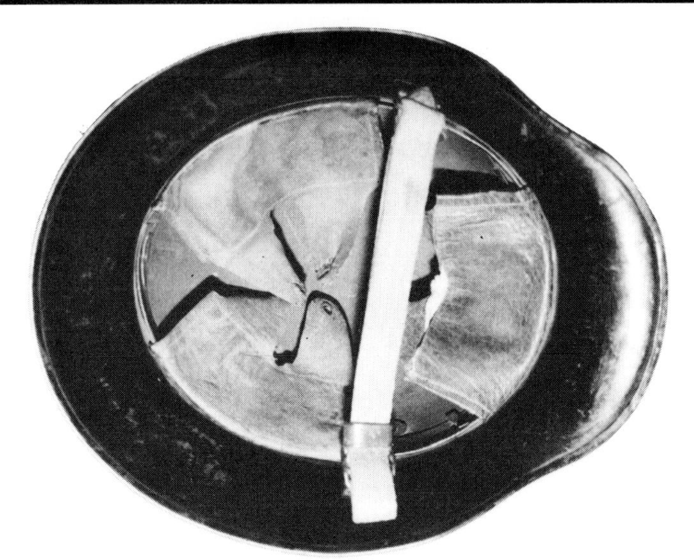

UPPER: *Interior view of Allgemeine-SS helmet displaying 1913 dated liner with mid-1930's repairs. The dome of this helmet retains the original field grey issue color.* **Hicks collection.**

LOWER: *Interior view of helmet showing the WWI Austrian liner and web chinstrap configuration. The unpainted brown dome is visible.* **Hicks.**

UPPER: Allgemeine-SS helmet with hand-painted satin black over Austrian Brown WWI finish. The decal is the classic 1st pattern shield with high metallic content and fine age toning. Hicks collection.

LOWER: Party shield side of same helmet, showing the typical pattern "thin swastika" shield. Hicks collection.

UPPER: 1st pattern SS runic shield showing slight age toning and high metallic content. **Hicks.**

LOWER: Party shield with "thick" swastika. This pattern can be discerned from the thinner variety. **Hicks.**

UPPER: Unlacquered 1st pattern shield which displays unmistakable brightness compated to lacquered decals. **Hicks collection**.

LOWER: Unlacquered 1st pattern shield on a rough texture finish. **Hicks.**

UPPER: Lacquered 2nd pattern shield on parade finish. The lacquer, which is seen going beyond the borders of the decal, has slightly toned the SS decal. Terry Goodapple collection.

LOWER: Standard 2nd pattern party shield. Note thicker border which matches the runic shield. Terry Goodapple collection.

2nd pattern runic shield on a rough-texture finish. The decals thinness allows it to "snuggle" into the paint surface. The lacquer coating is visible outside the borders. **Hicks collection**.

UPPER: *Variant, probably private purchase SS decal. Note the high metallic content, blocky runes, and lacquer coating.* **Chris Jones**.

LOWER: *Standard pattern swastika shield is used in conjunction with the variant runic shield.* **Chris Jones collection.**

UPPER: *Austrian manufactured variant 1st pattern runic shield. Note the thicker border and closeness of the runes.* **Hicks collection**.

LOWER: *Variant swastika shield on Austrian SS-VT helmet. Note the thicker border and large swastika circle.* **Hicks collection**.

Mirror runic shield and variant swastika shield, both of Austrian manufacture. The runic shield is believed to be a misprint by a local manufacturer. The party shield is typical of Austrian pre-war designs. **Al Barrows collection.**

M-35 Waffen-SS helmet with light tan and brown camouflage finish. The runes are the 2nd pattern. **Al Barrows collection.**

Swastika shield side of the same camouflage helmet shown on the facing page. **Al Barrows collection.**

M-35 SS police helmet consisting of standard combat police helmet with the addition of a 1st pattern SS runic shield atop the unbordered police decal. **Al Barrows collection.**

Right side view of SS police helmet. Many foreign volunteer units of the SS were initially police organizations. **Al Barrows collection**.

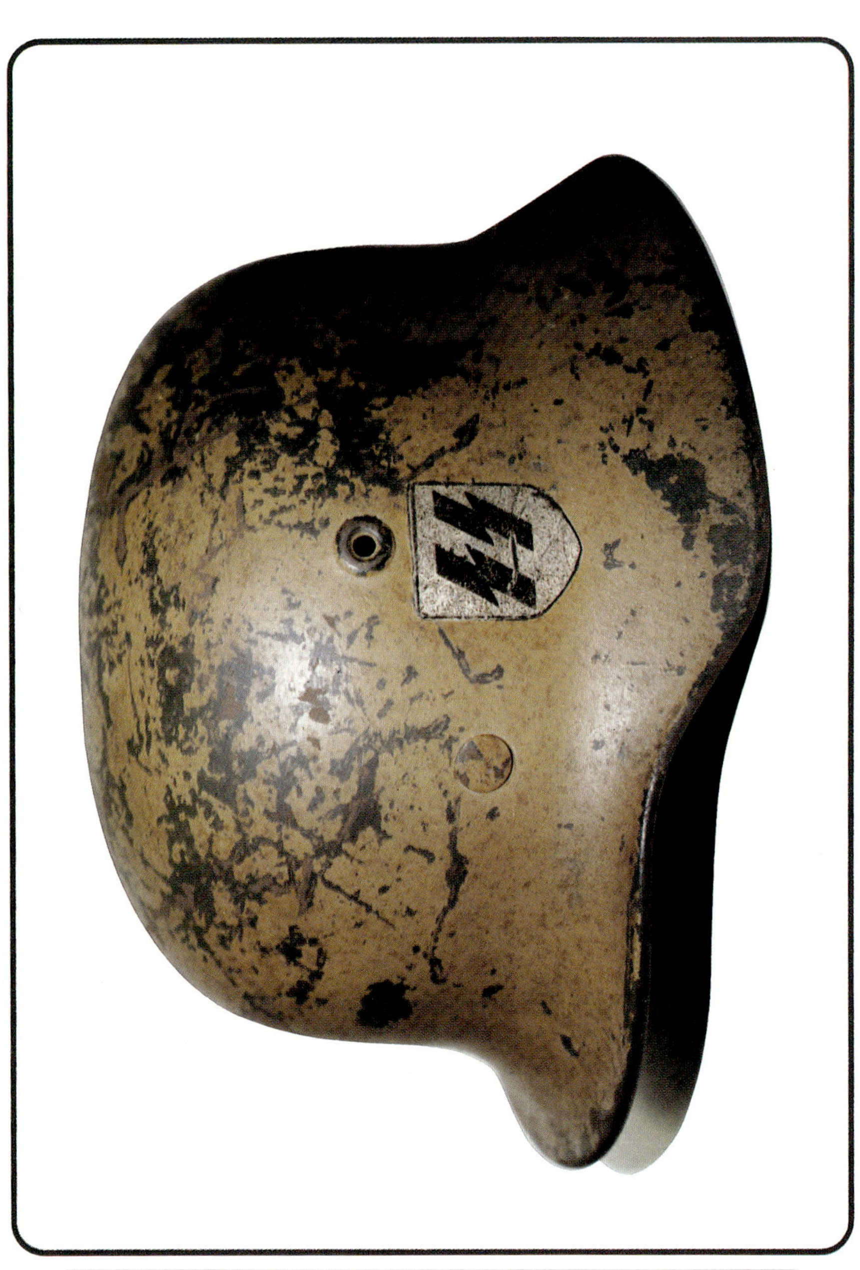

Unusual double decal M-40 tropical camouflaged helmet with 1st pattern runic shield placed directly atop a bordered Schutzpolizei decal. This helmet is probably an example of the procedural insignia change which took place when police units were reassigned to the SS. **Al Barrows collection**.

Right side of the SS/Police helmet shown on the facing page. Note the proper configuration for the police swastika shield. **Al Barrows collection**.

SS Two views of a strikingly-camouflaged field police helmet with the unbordered police eagle characteristic of SS police units. **Russ Hamilton collection**.

Civic style SS double decal helmet with 1st pattern lacquered shield. This helmet is marked in a manner which indicates a non-combatant or guard role. *Al Barrows collection*.

Left side of civic style SS helmet shown on the previous page.
Al Barrows Collection.

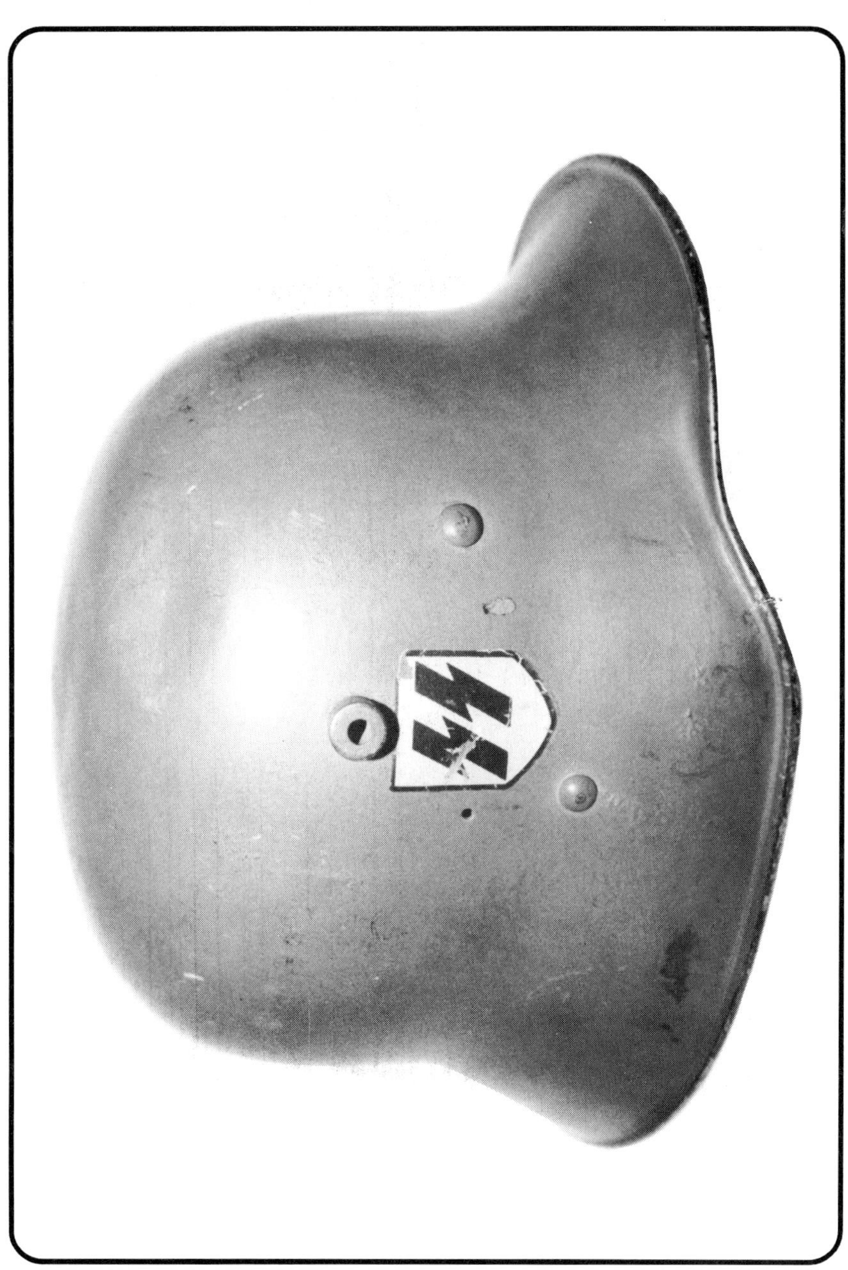

Runic side view of SS-VT helmet with runes believed to be of pre-war Austrian manufacture. The finish is earth brown over satin black over field grey. Also note the WWI insignia apertures. **Hicks collection**.

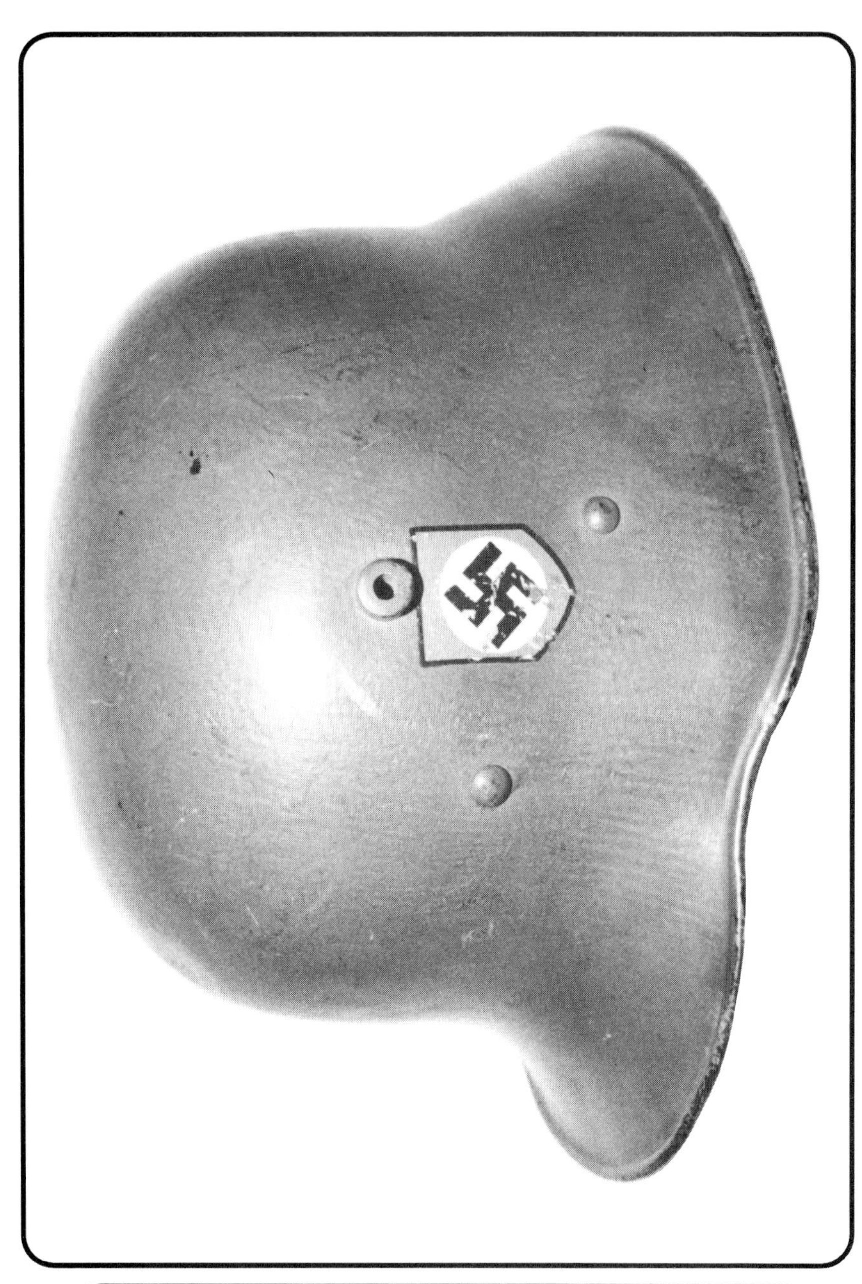

Left view of helmet clearly showing the variant party shield, which can be found in Goodapple, Vol 2, page 143. **Hicks Collection.**

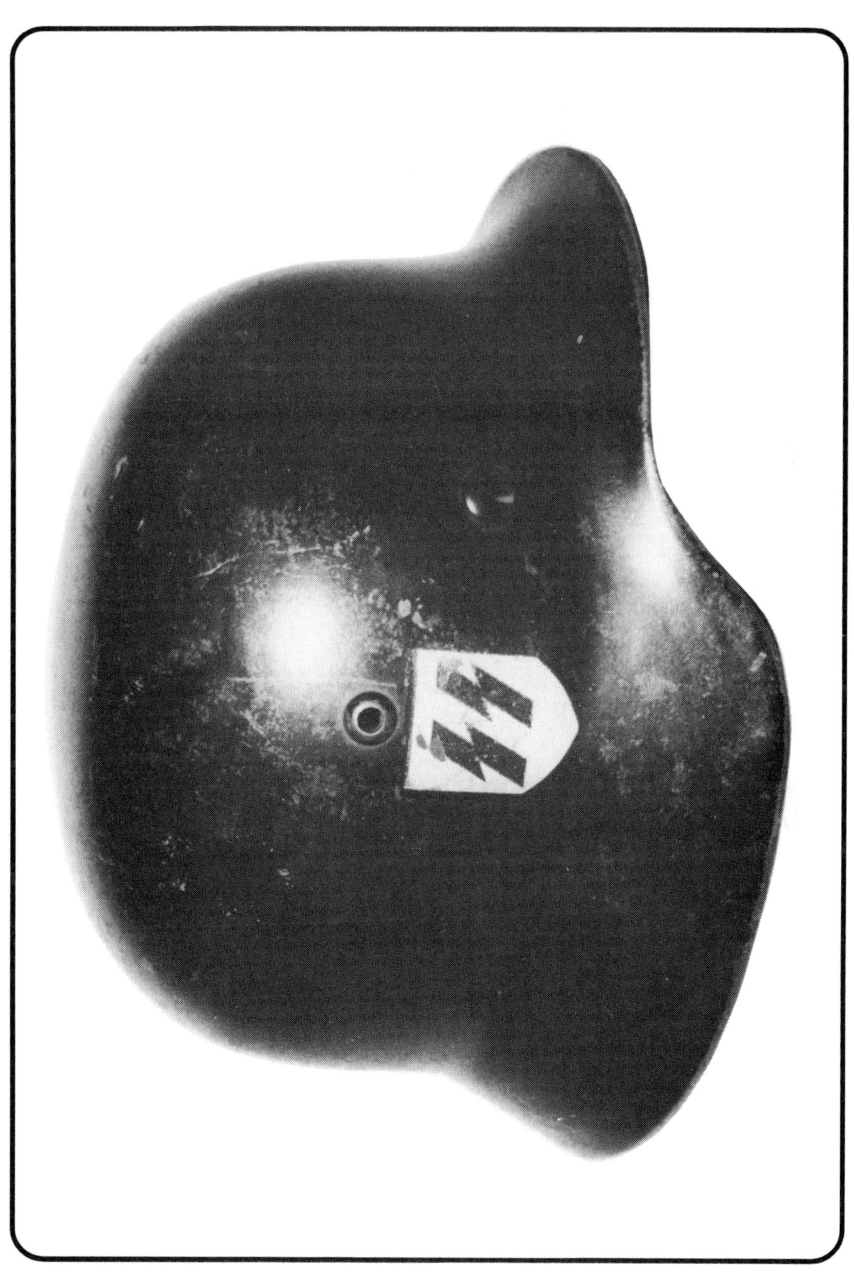

Allgemeine-SS M-35 helmet with satin black over field grey finish. This helmet, as are most of its type, was originally a field grey double decal SS helmet. **Hicks collection**.

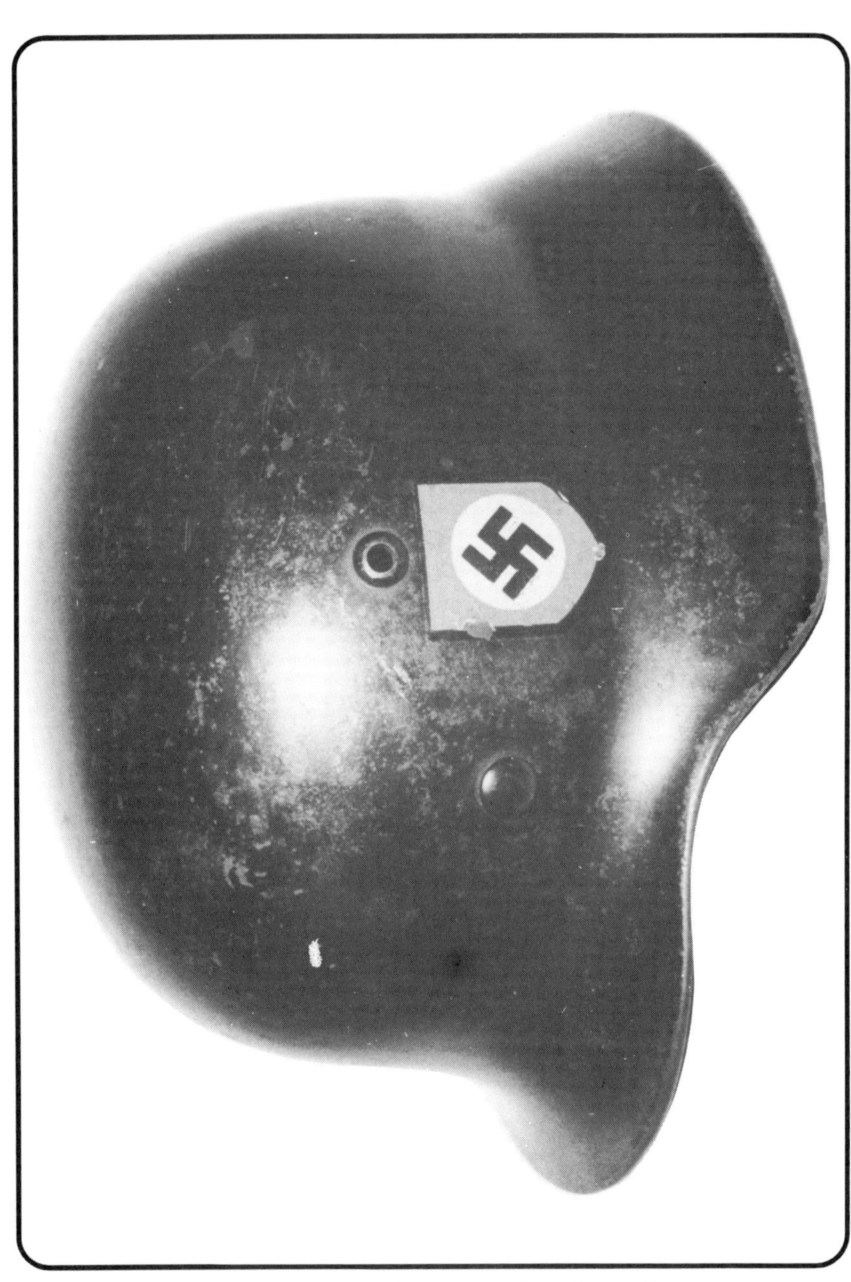

The "thick" variety swastika shield is clearly visible here. This helmet is dated 1939. **Hicks collection.**

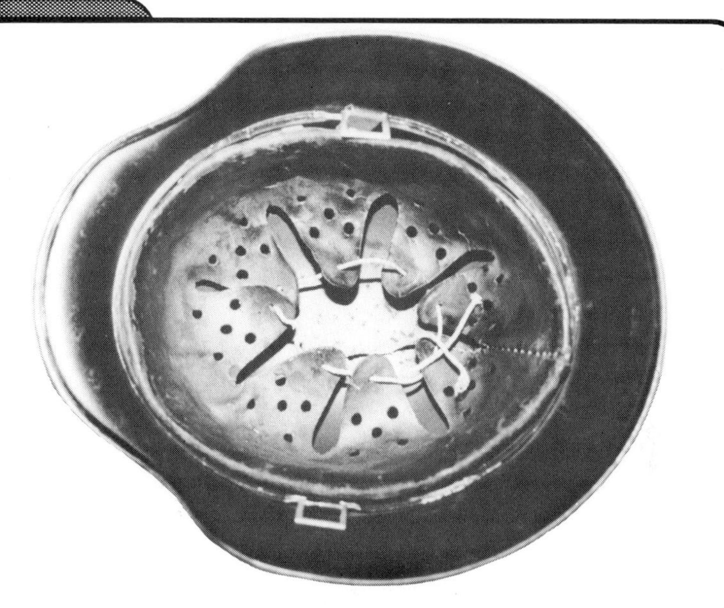

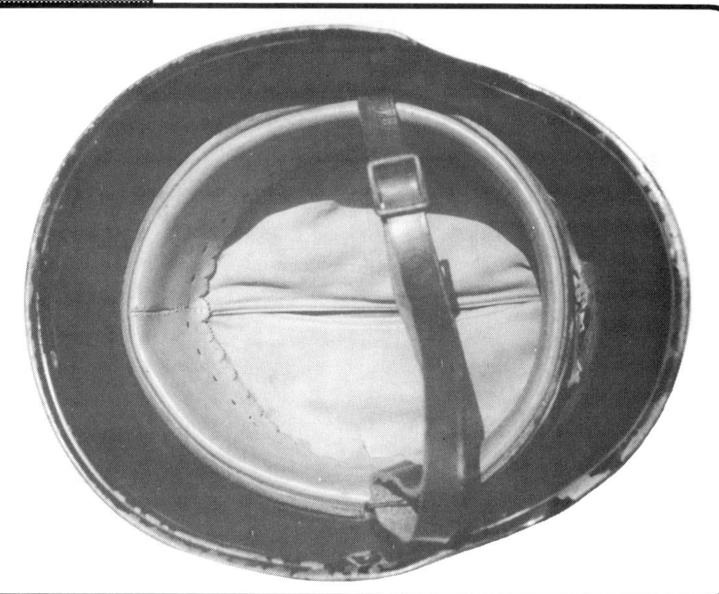

UPPER: *Interior view of M-35 helmet. Again, note the unpainted field grey dome.* **Hicks collection**.

LOWER: *The interior of the parade M-35 shows a commercially produced liner.* **Al Barrows collection**.

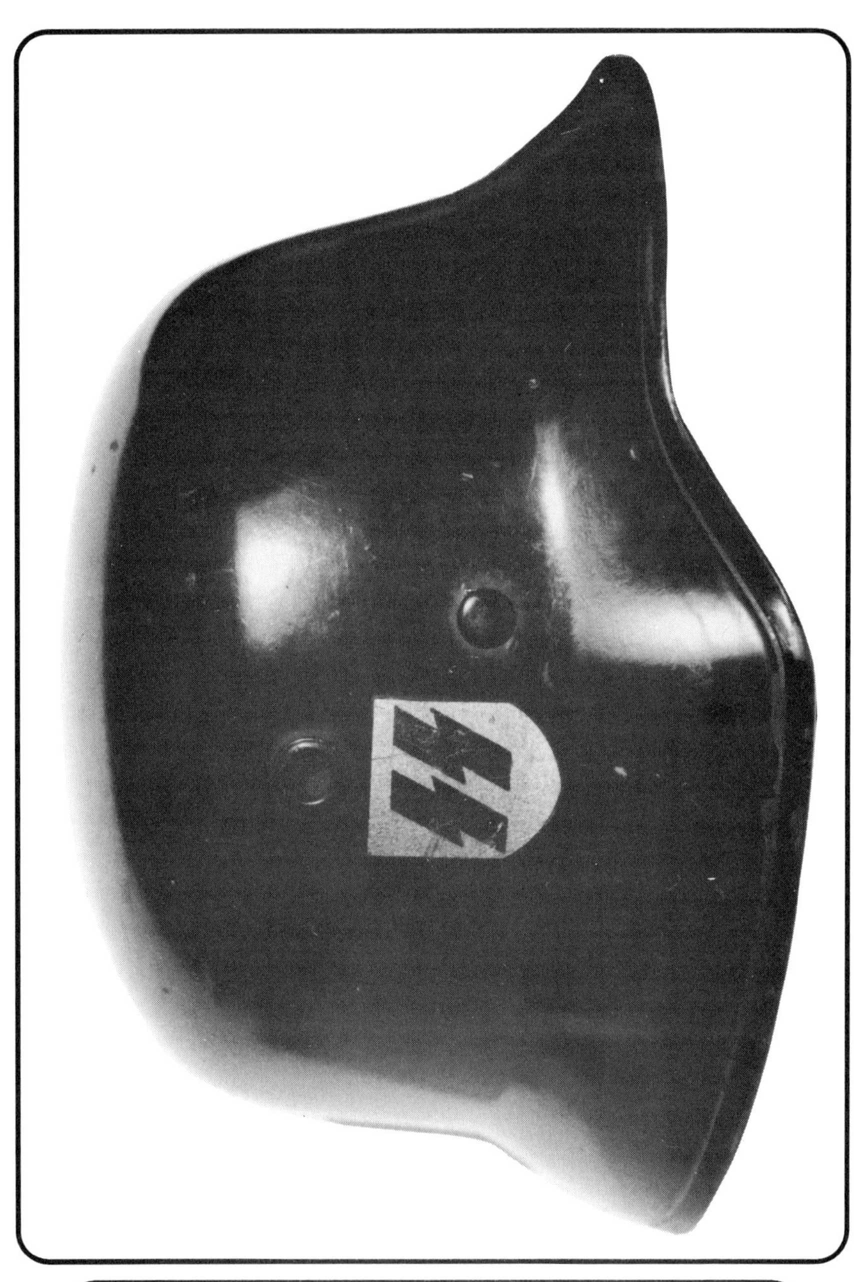

Rare vulkanfiber M-35 style parade helmet, showing variant runic shield. **Al Barrows collection.**

*Party shield side of parade helmet shown on left. Note extensive age cracking to the decal. **Al Barrows collection**.*

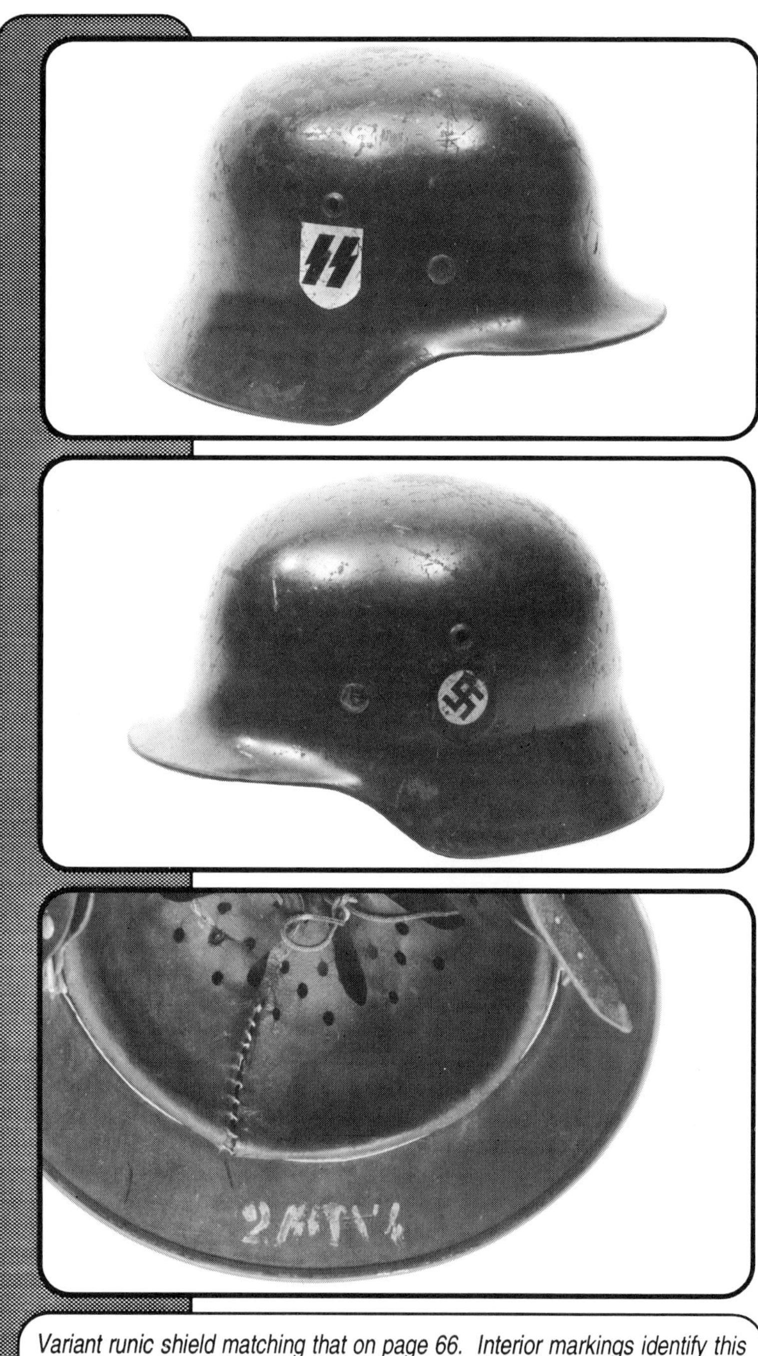

Variant runic shield matching that on page 66. Interior markings identify this M-35 to the SS-TV regiment "Ostmark." **Chris Jones collection**.

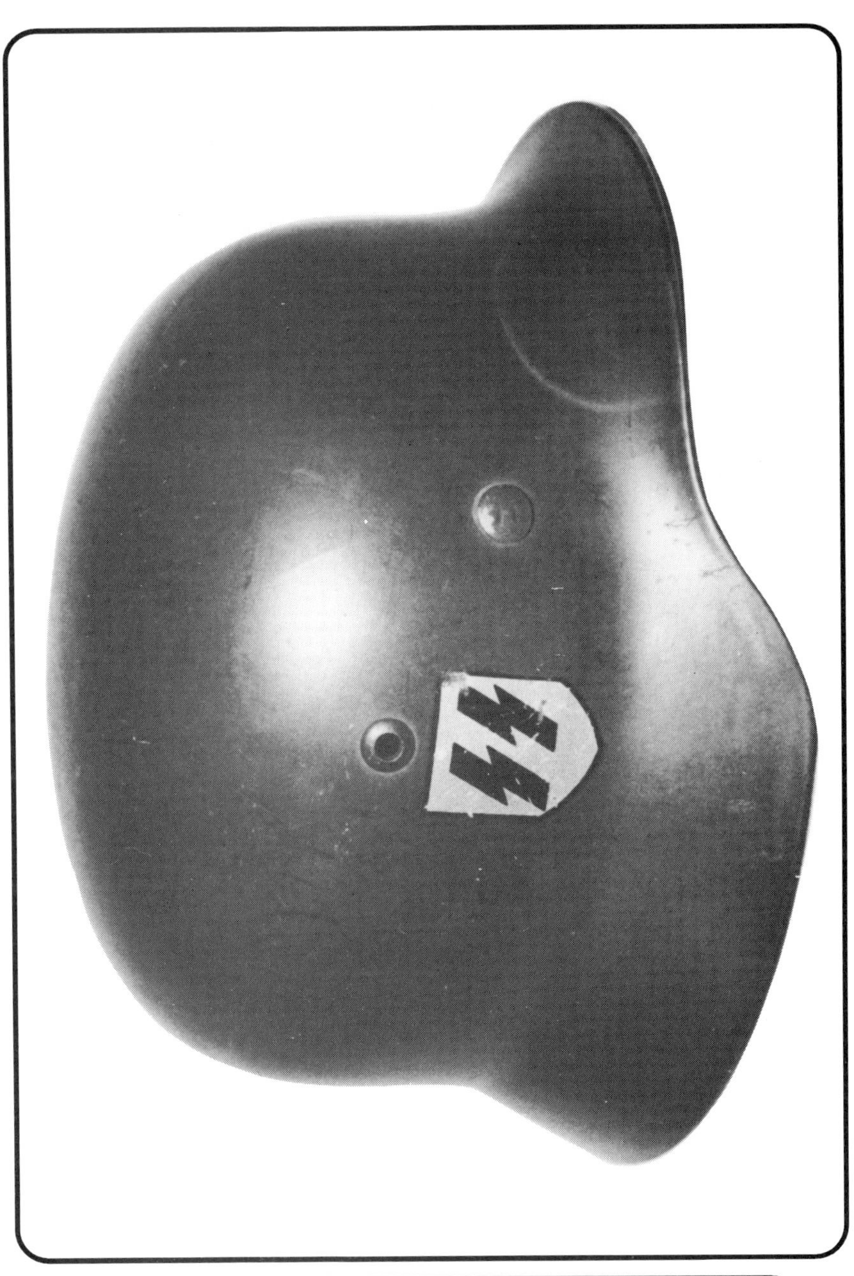

Classic M-35 "Waffen-SS: helmet with medium green parade finish and lacquered 1st pattern runes. **Hicks collection**.

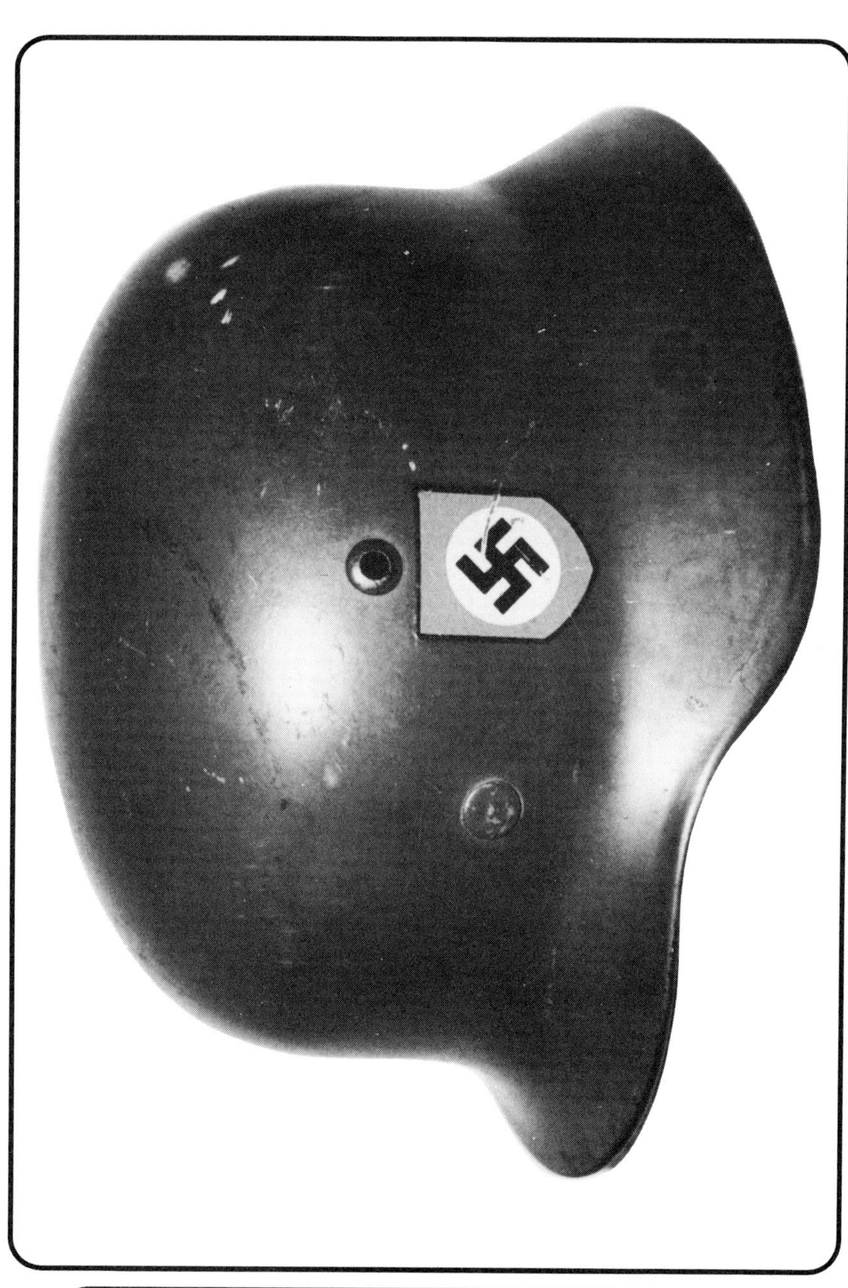

Left side of the helmet shown on page 69. Note the age cracks within the swastika circle, as well as the manner in which the lacquer coating extends beyond the decals borders. **Hicks collection**.

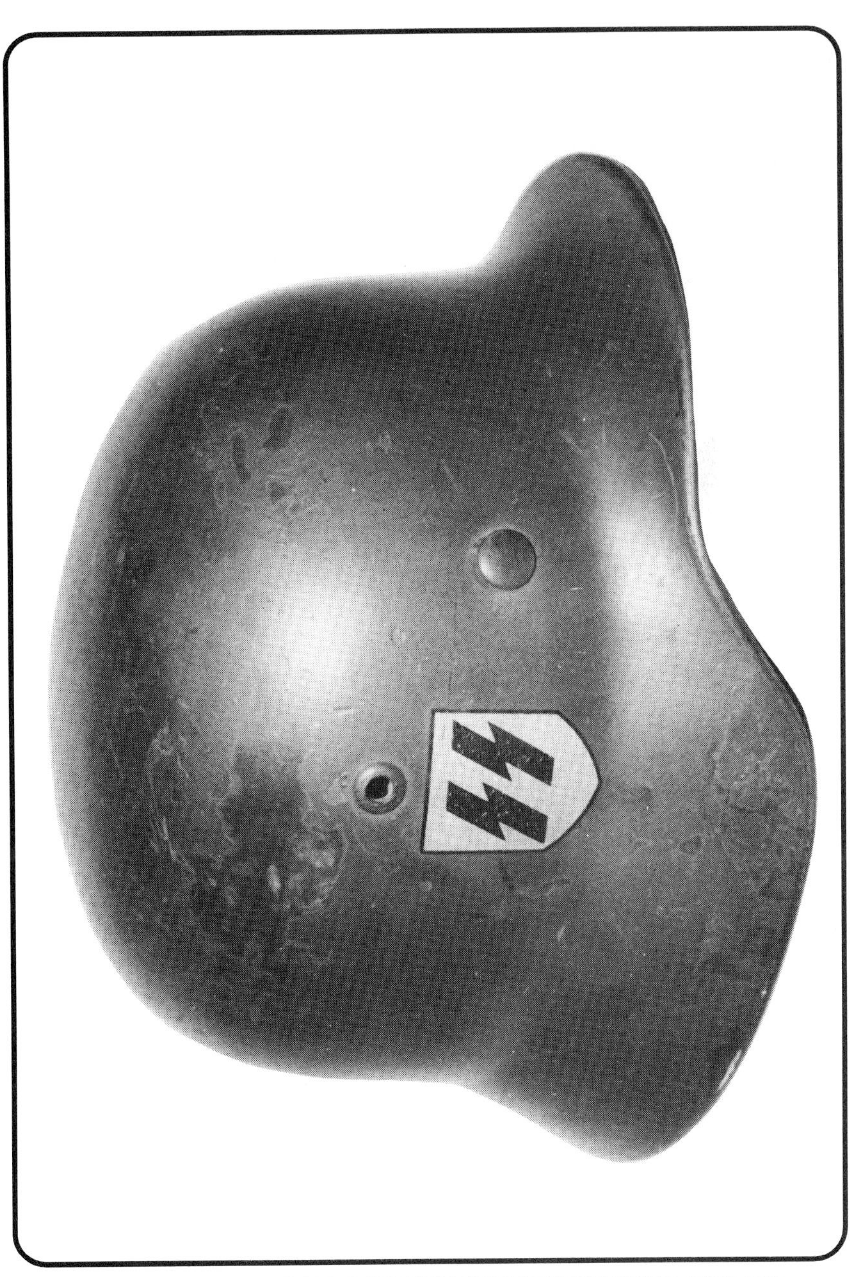

M-35 Waffen-SS re-issue helmet with medium grey finish over parade green.
Hicks collection.

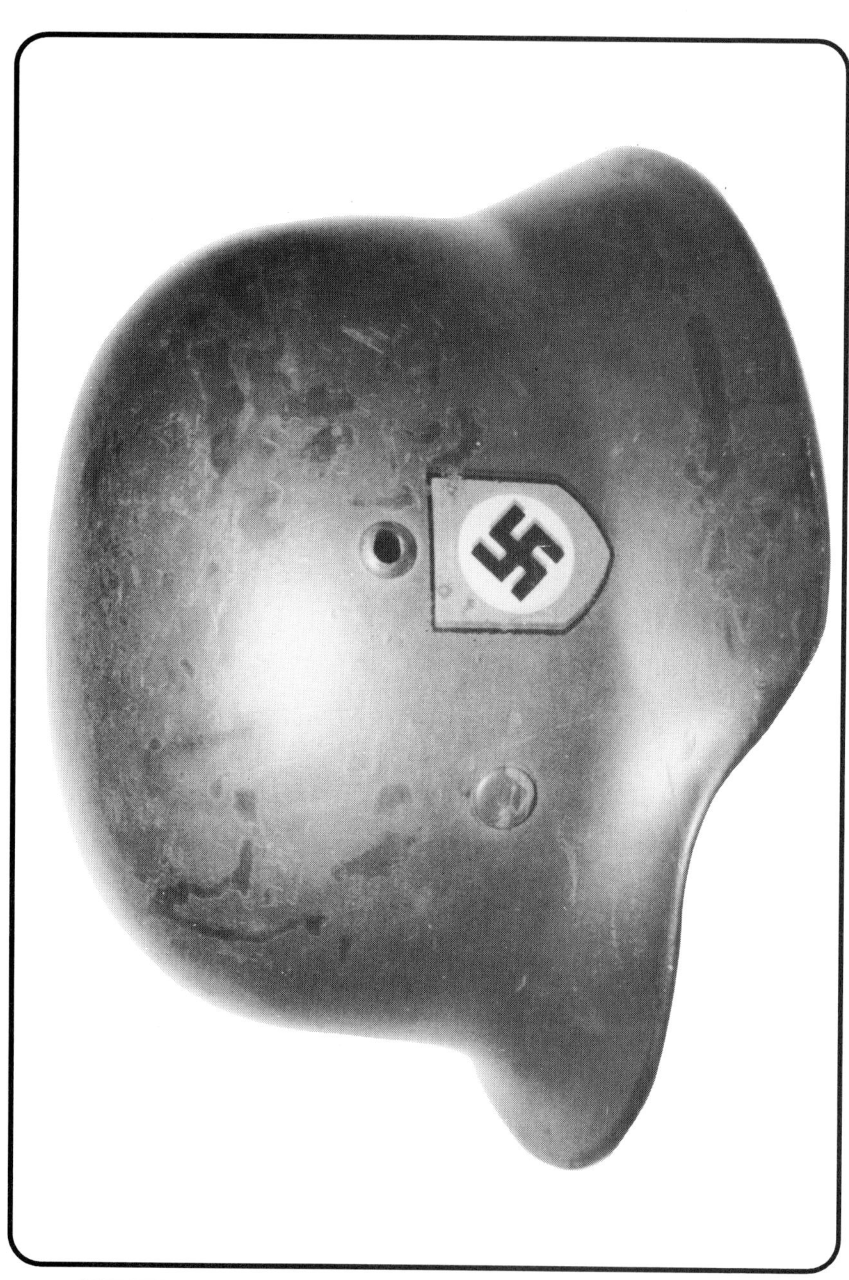

Left side of the helmet on the previous page, reveals another example of the "thick style swastika. *Hicks collection*.

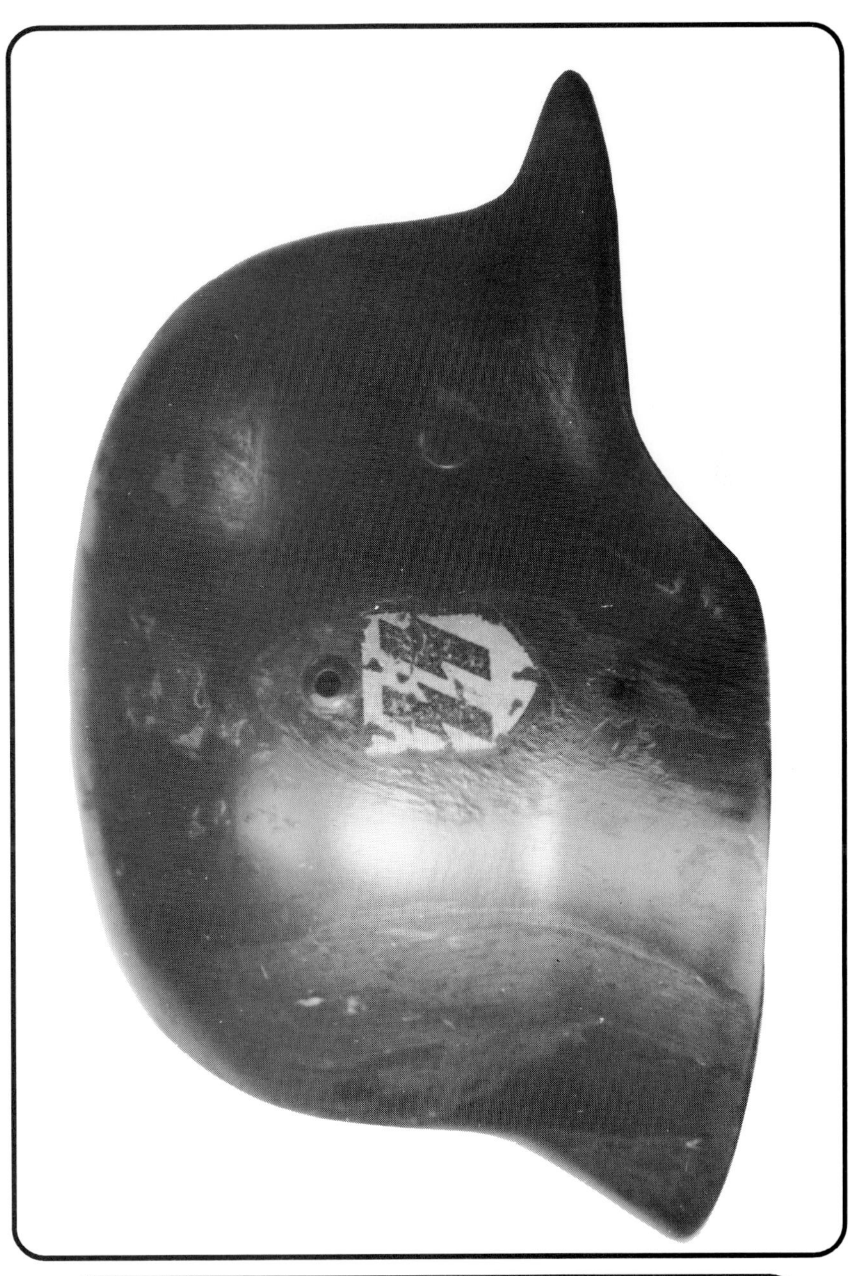

Unique example of camouflaged M-35 Waffen-SS helmet, showing a medium green paint over geometric whitewash patterns. The helmet is dated 1938. **K.R. Bowra collection**

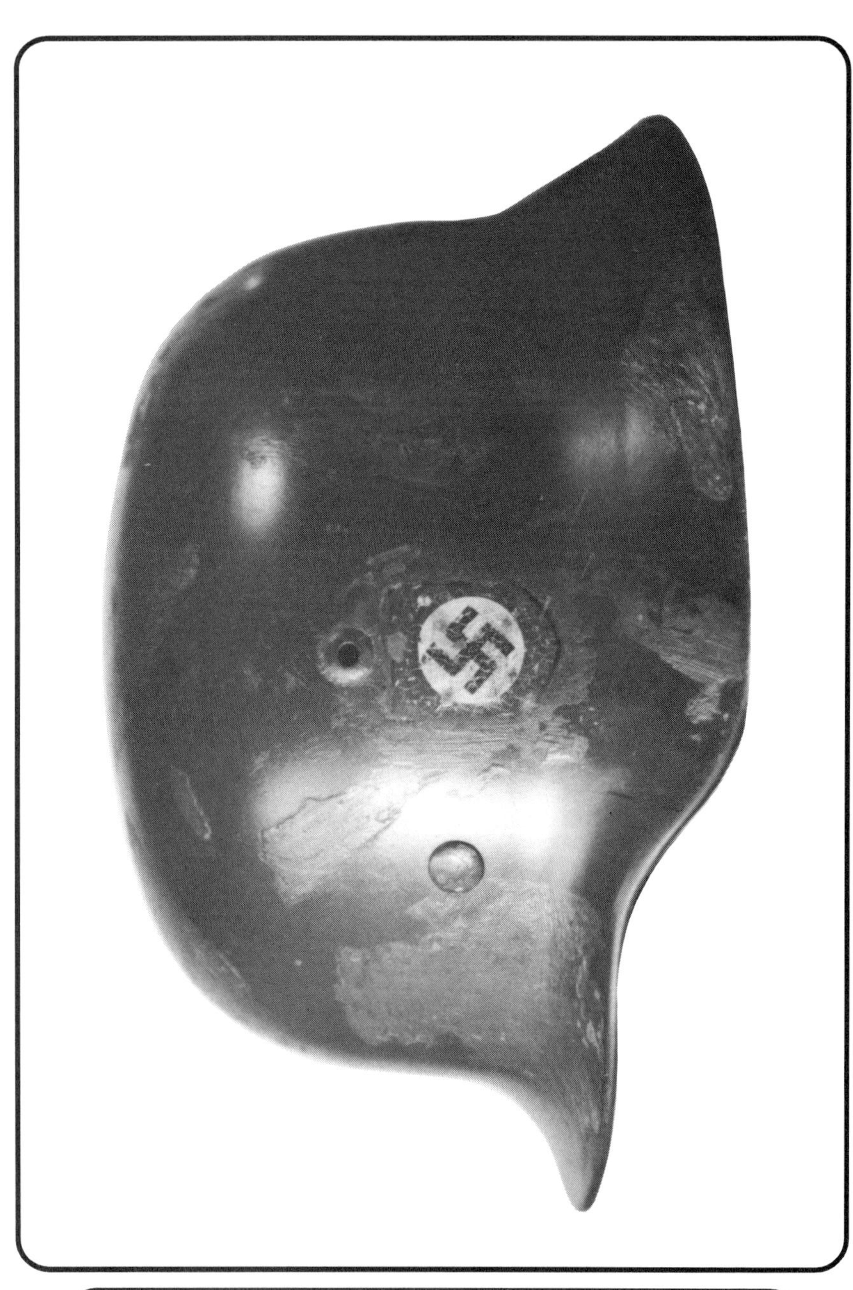

Note the age-cracking to the swastika shield on this left side view of the 1938 dated M-35 Waffen-SS helmet. **K.R. Bowra collection**.

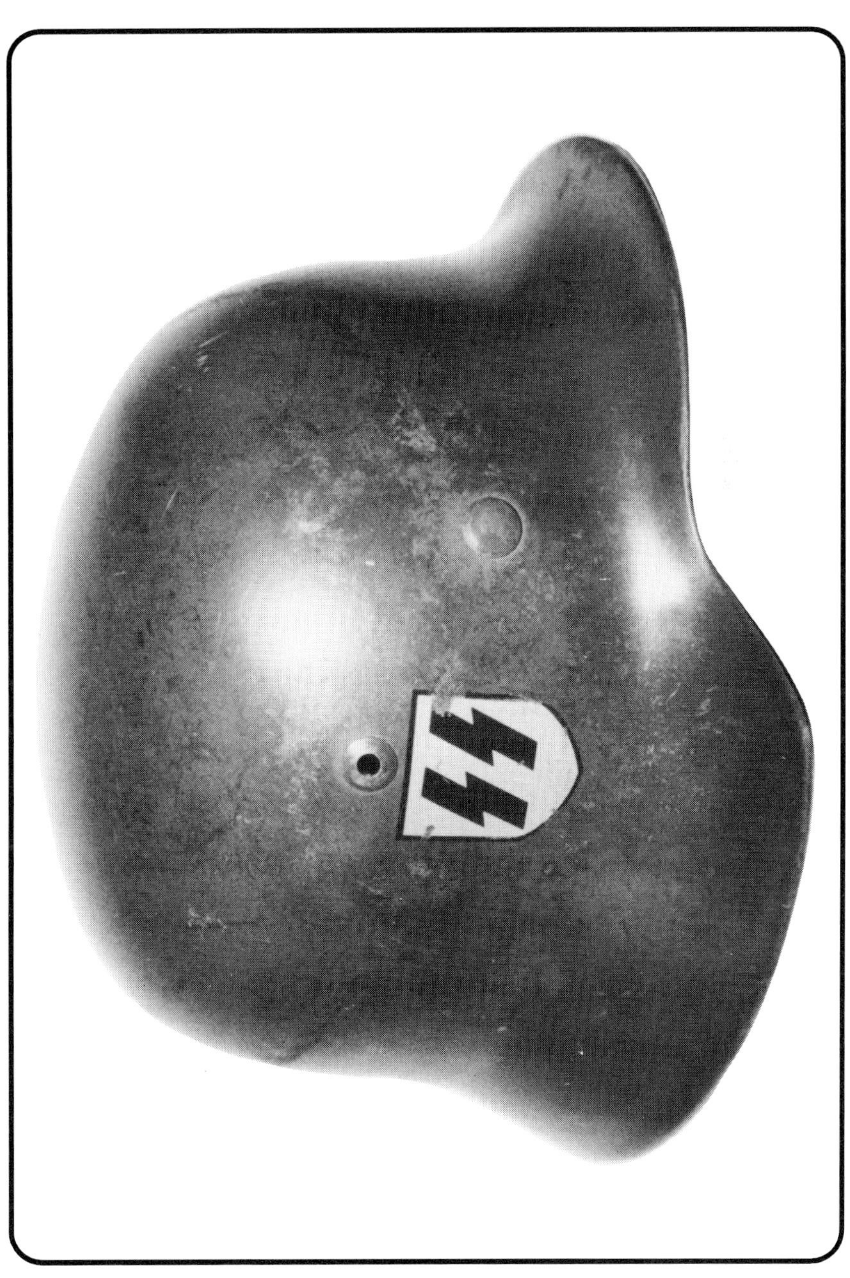

*Waffen-SS M-35 helmet with glossy parade finish dark green paint and 2nd pattern shield. This helmet is dated 1937. **Hicks collection**.*

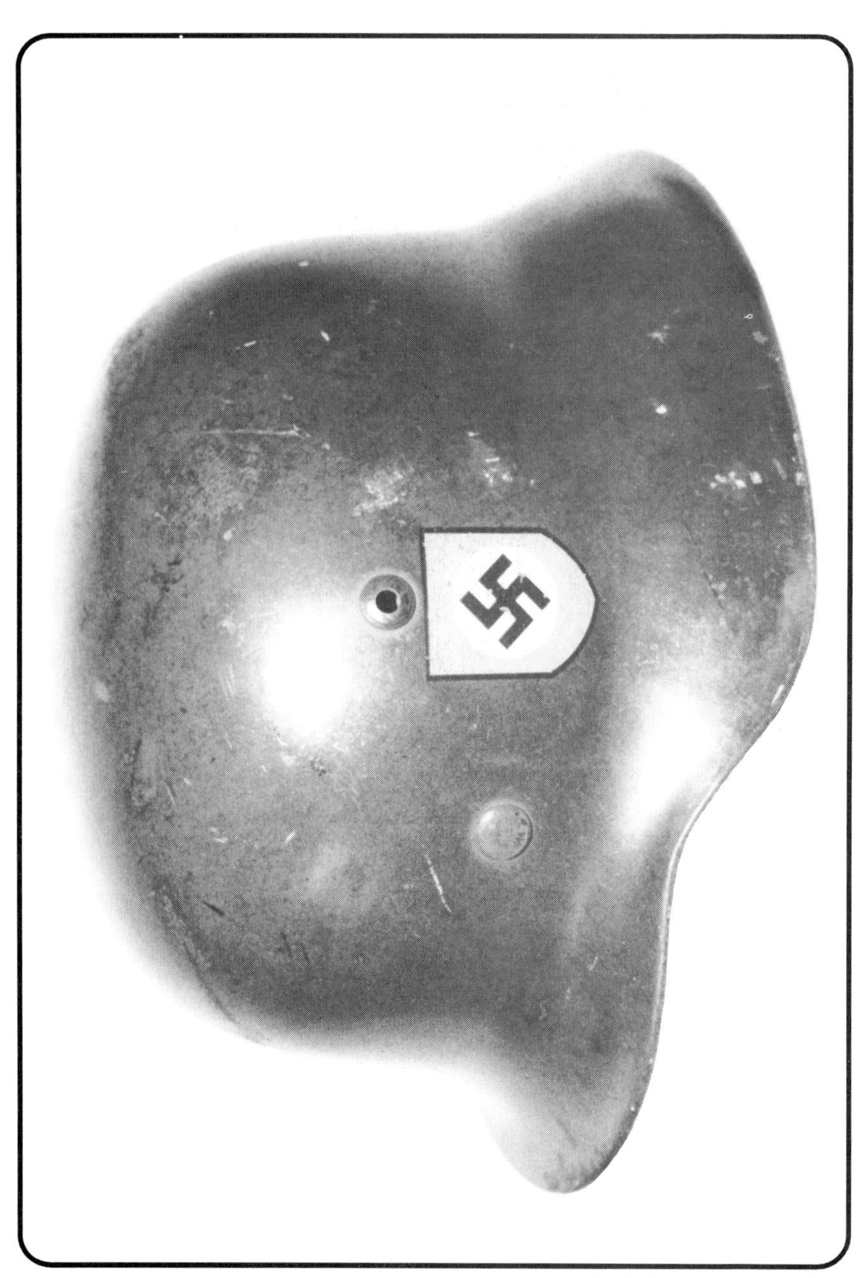

Standard pattern swastika shield for the 2nd pattern runes, consisting of thin style swastika, thicker outer border, and commonly off-center swastika circle. **Hicks collection.**

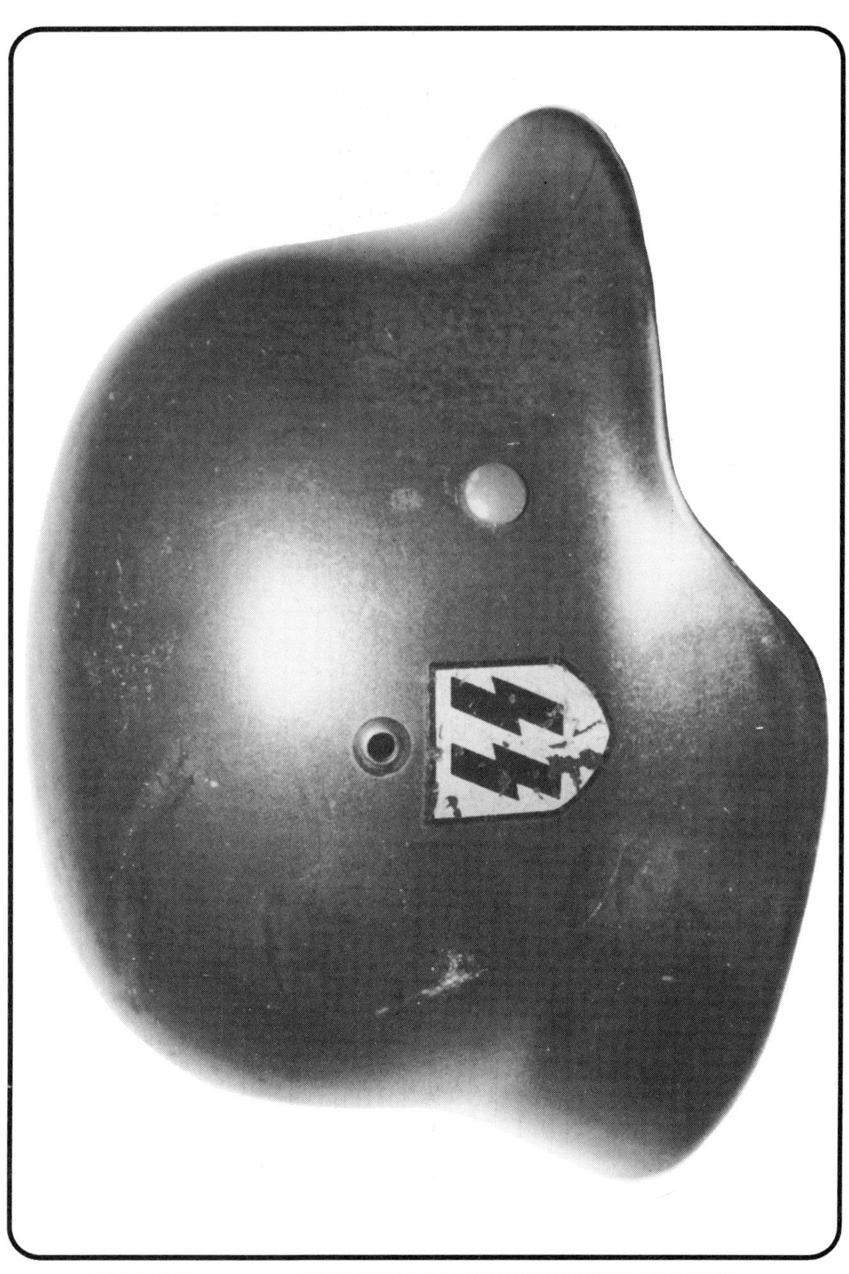

M-35 Waffen-SS helmet with lacquered 2nd pattern runes. The lacquering enabled the decals to survive a high degree of combat exposure. This helmet is dated 1938. **Hicks collection**.

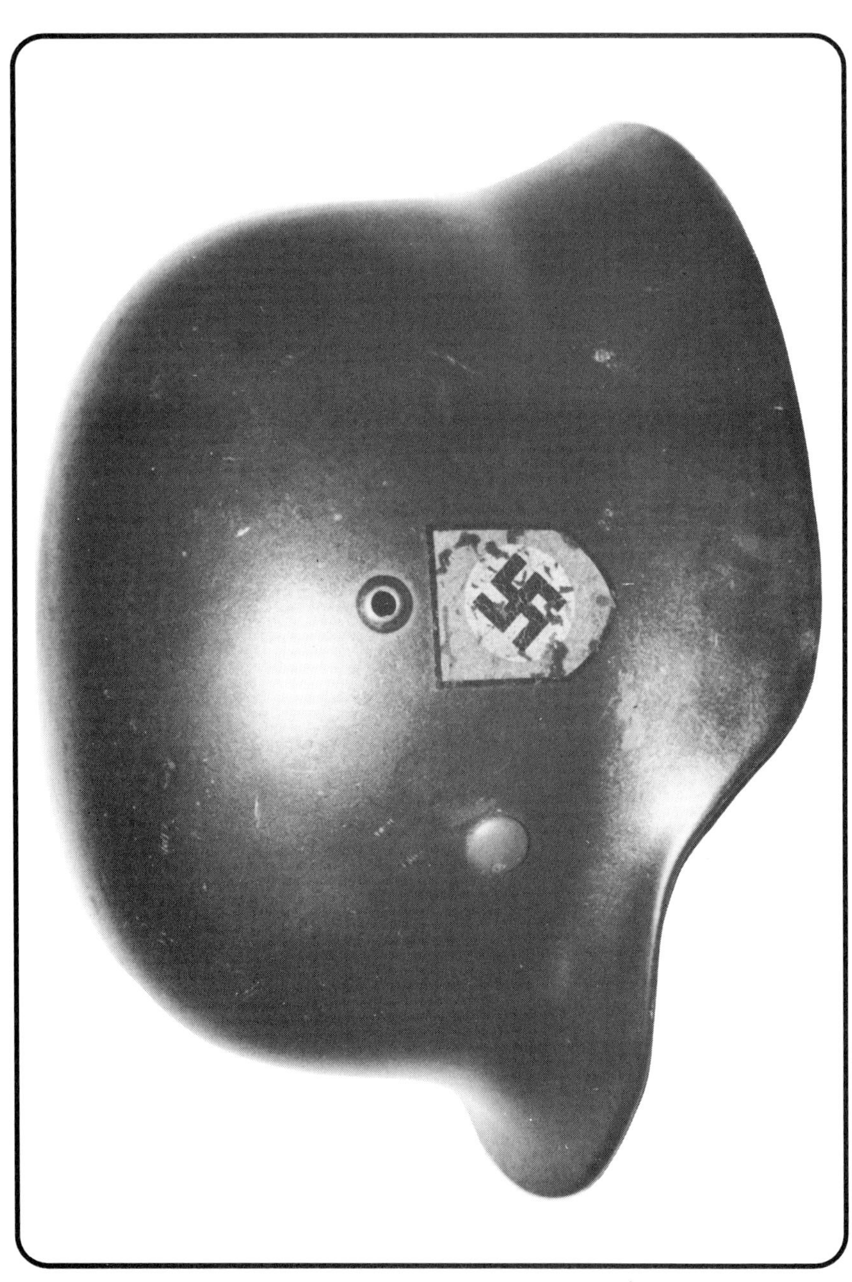

The high degree of age-cracking on this swastika shield attests to its heavy use. **Hicks collection**.

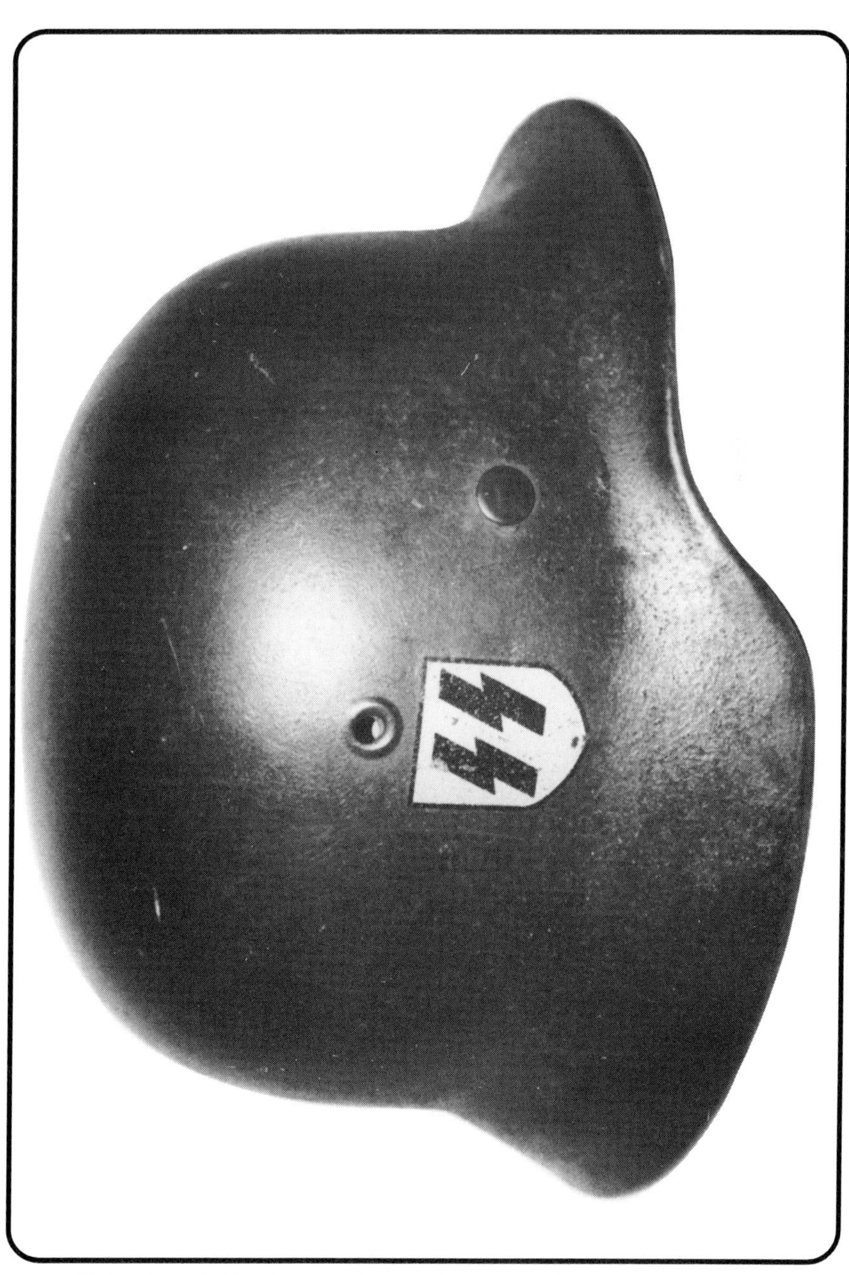

M-40 Waffen-SS double decal helmet with 2nd pattern runic shield and rough texture, field grey finish. This helmet is dated 1940. **Hicks**.

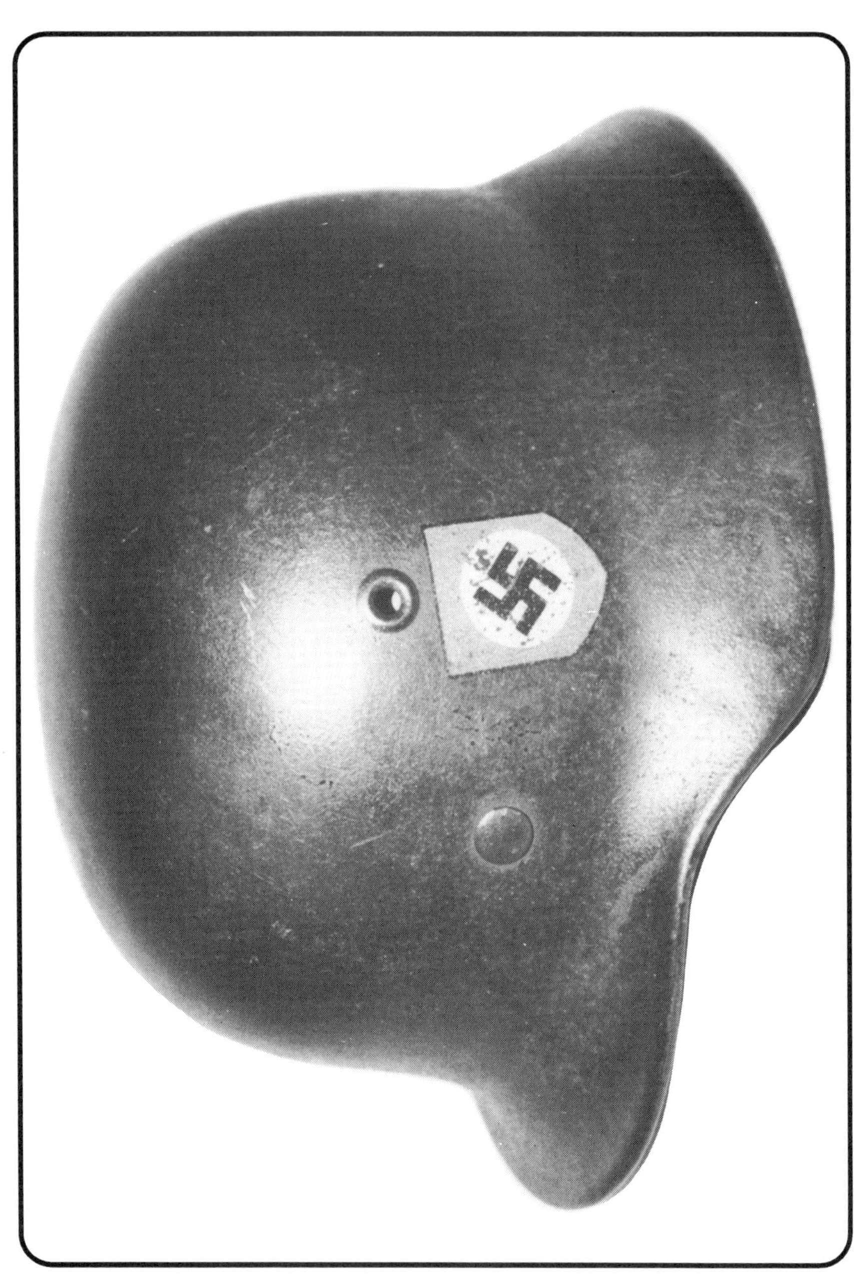

Left side view of M-40 double decal SS helmet showing the "thick" style swastika shield. **Hicks collection**.

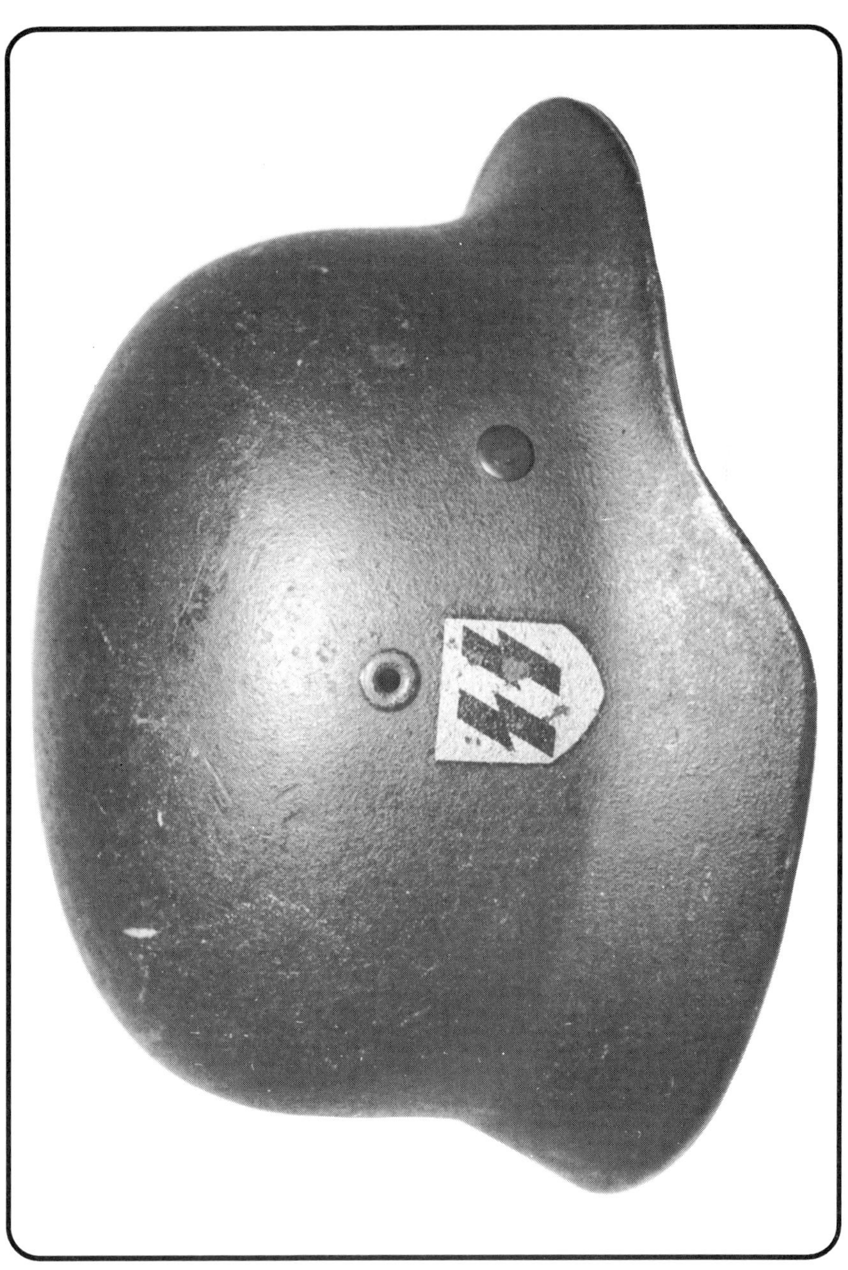

M-40 Waffen-SS single decal helmet with very rough texture finish. The 1st pattern runic shield is heavily gold toned. The helmet date is 1941. **Hicks collection**.

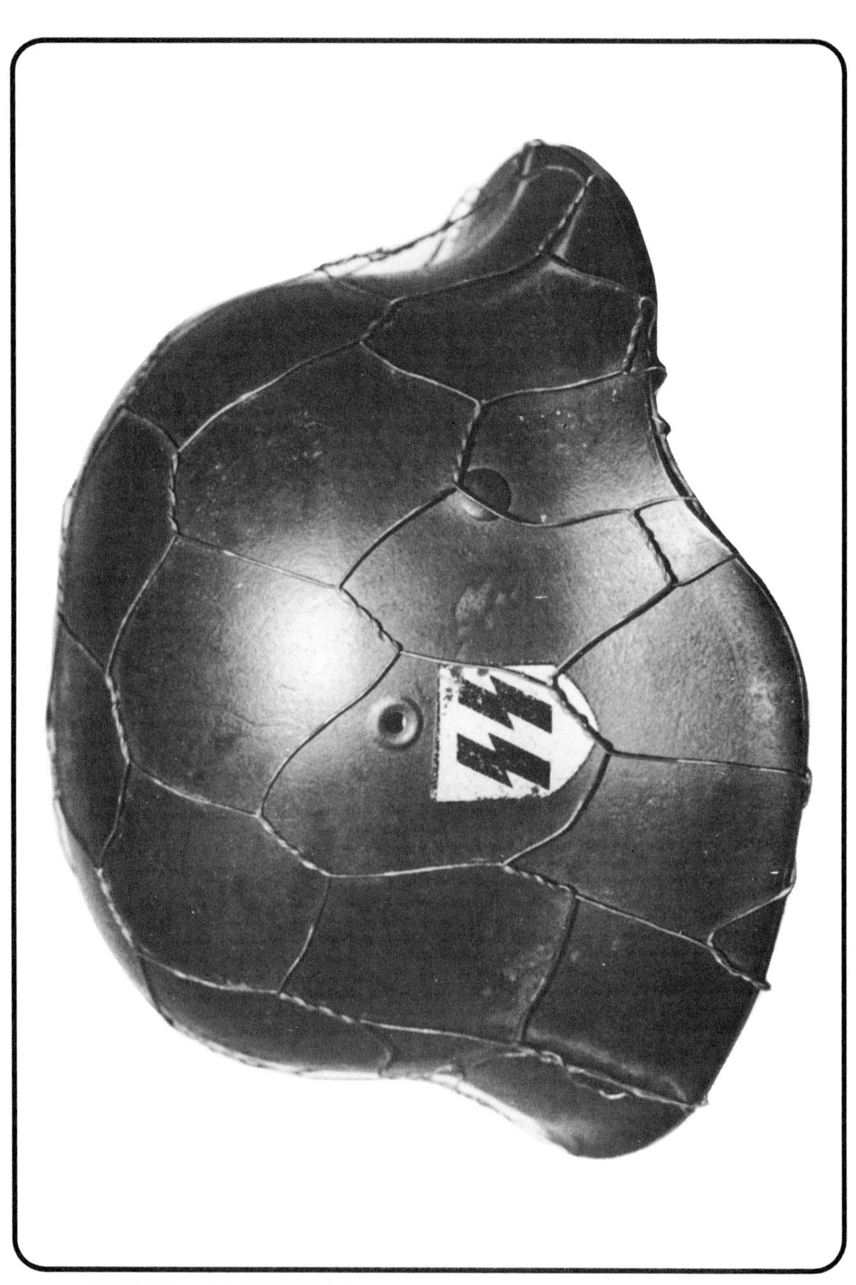

M-42 single decal Waffen-SS helmet with smoother textured medium grey paint, 1st pattern runic shield, and chicken wire cover for camouflage purposes. *Hicks collection*.

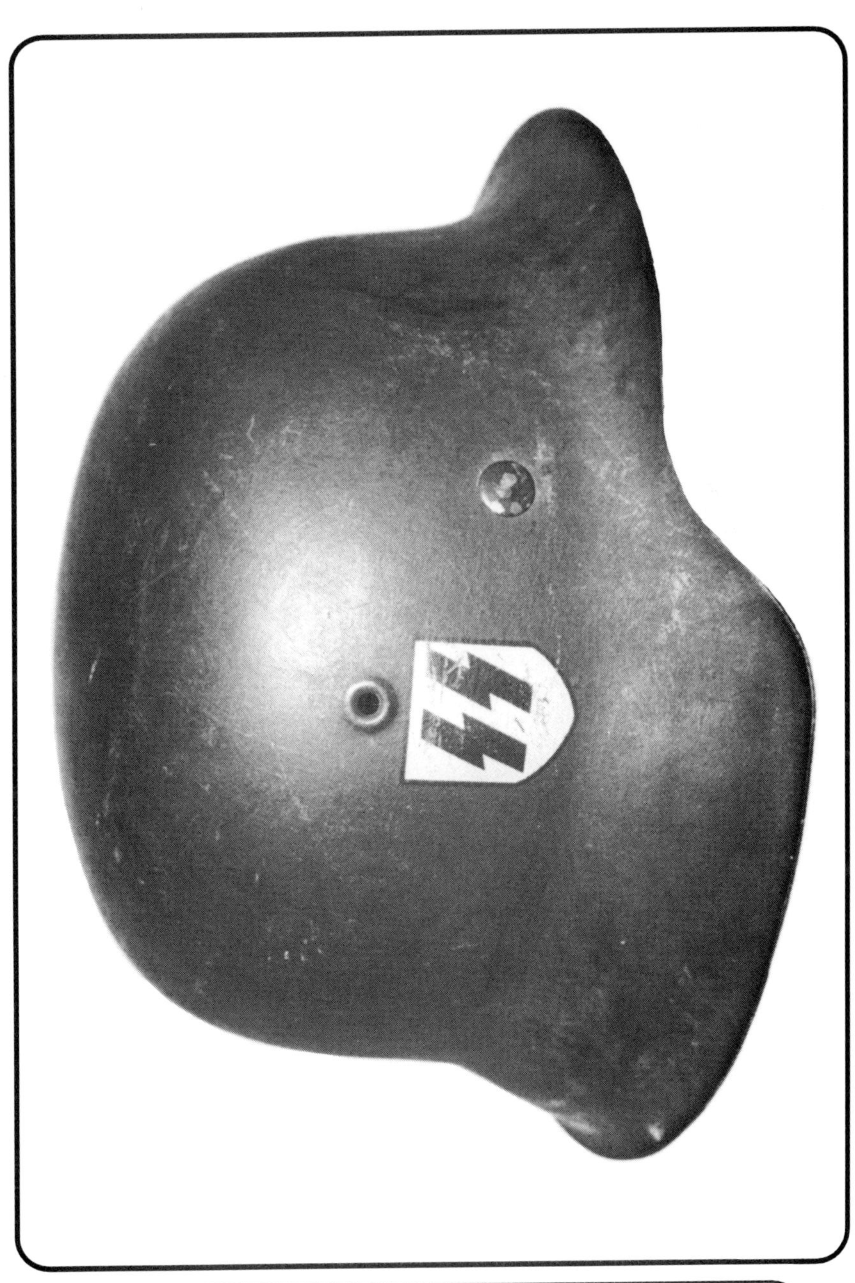

M-42 Waffen-SS single decal helmet with 2nd pattern runes and rough texture, medium grey finish. Hicks collection.

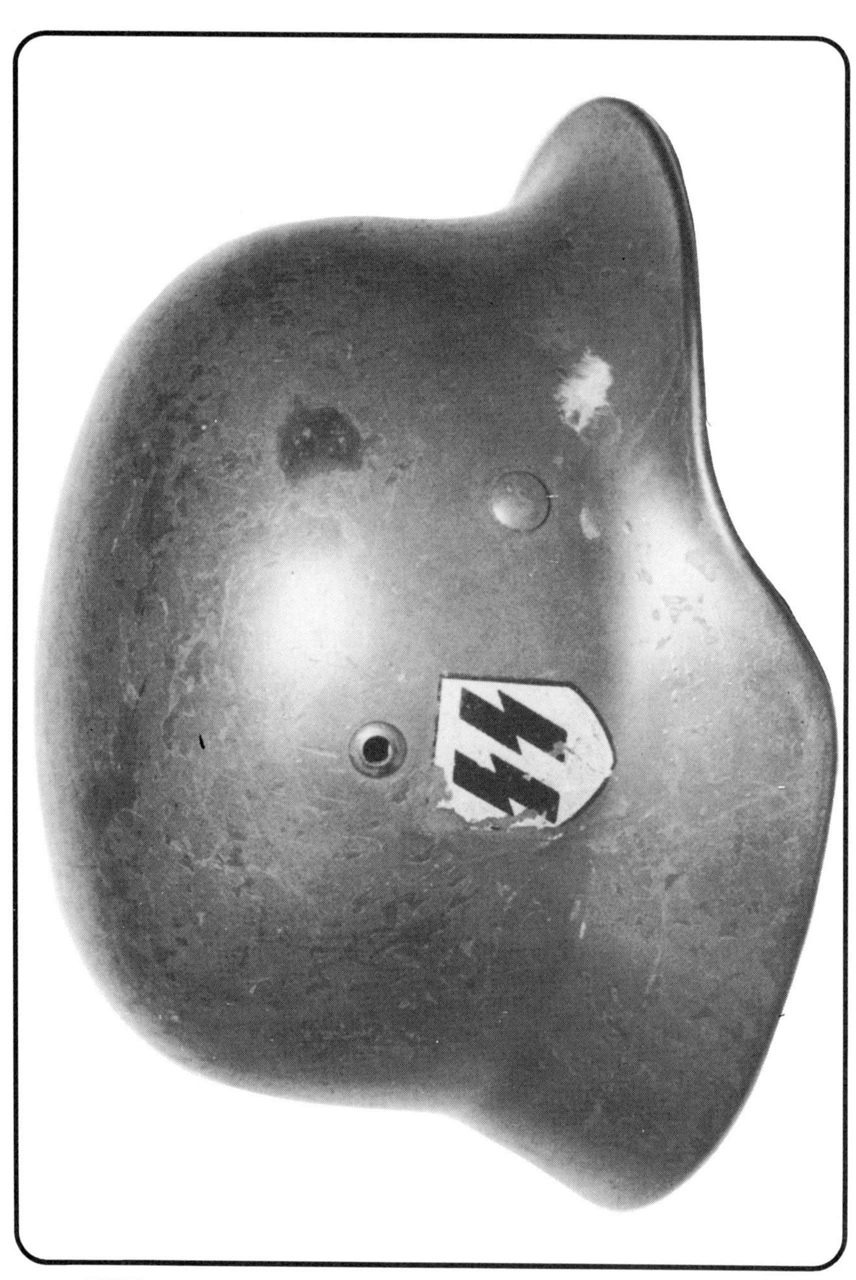

M-35 SS double runic helmet showing 1st pattern shield on repainted medium green finish. **Hicks collection**.

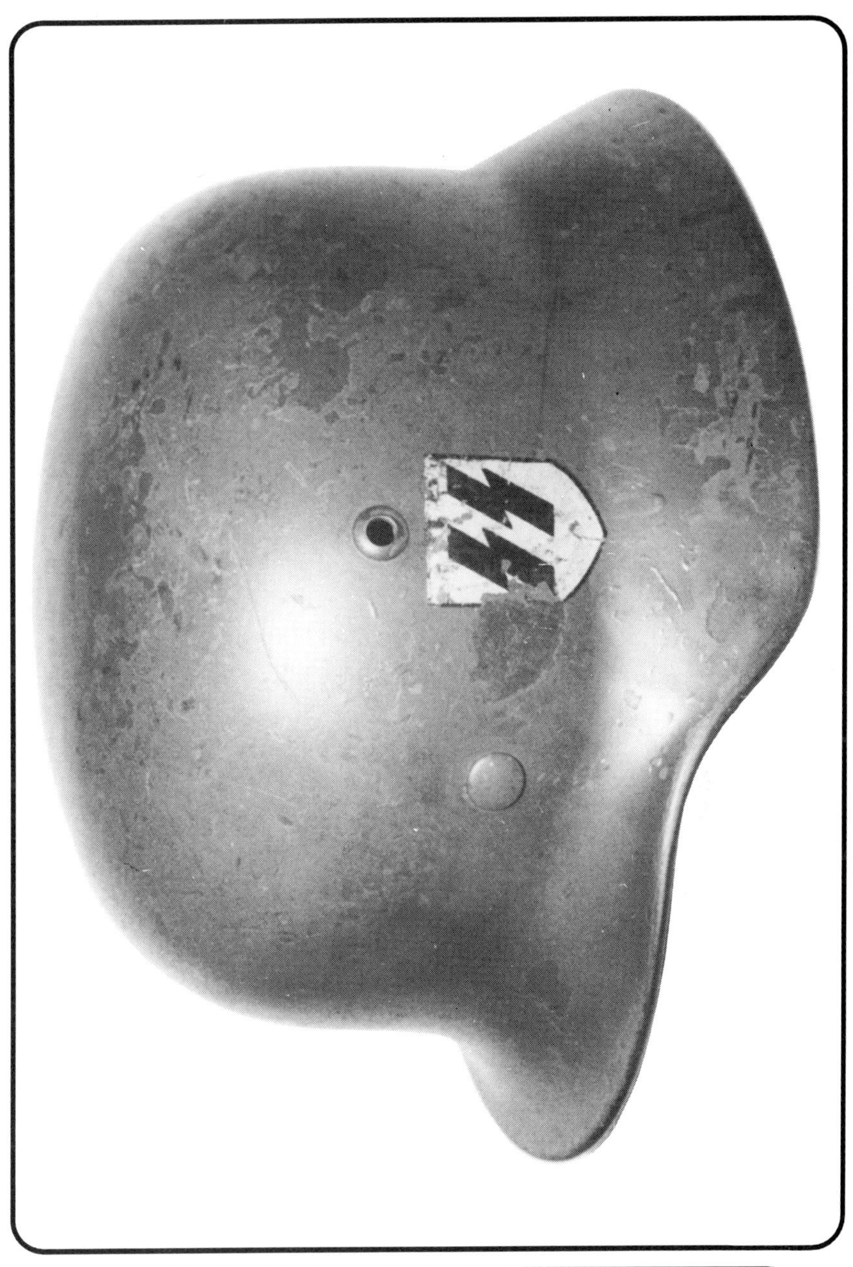

Left side of helmet showing runic shield atop refinished surface. These helmets are believed to represent either foreign volunteer use or simply individual decorative expression for esprit de corps purposes. **Hicks**.

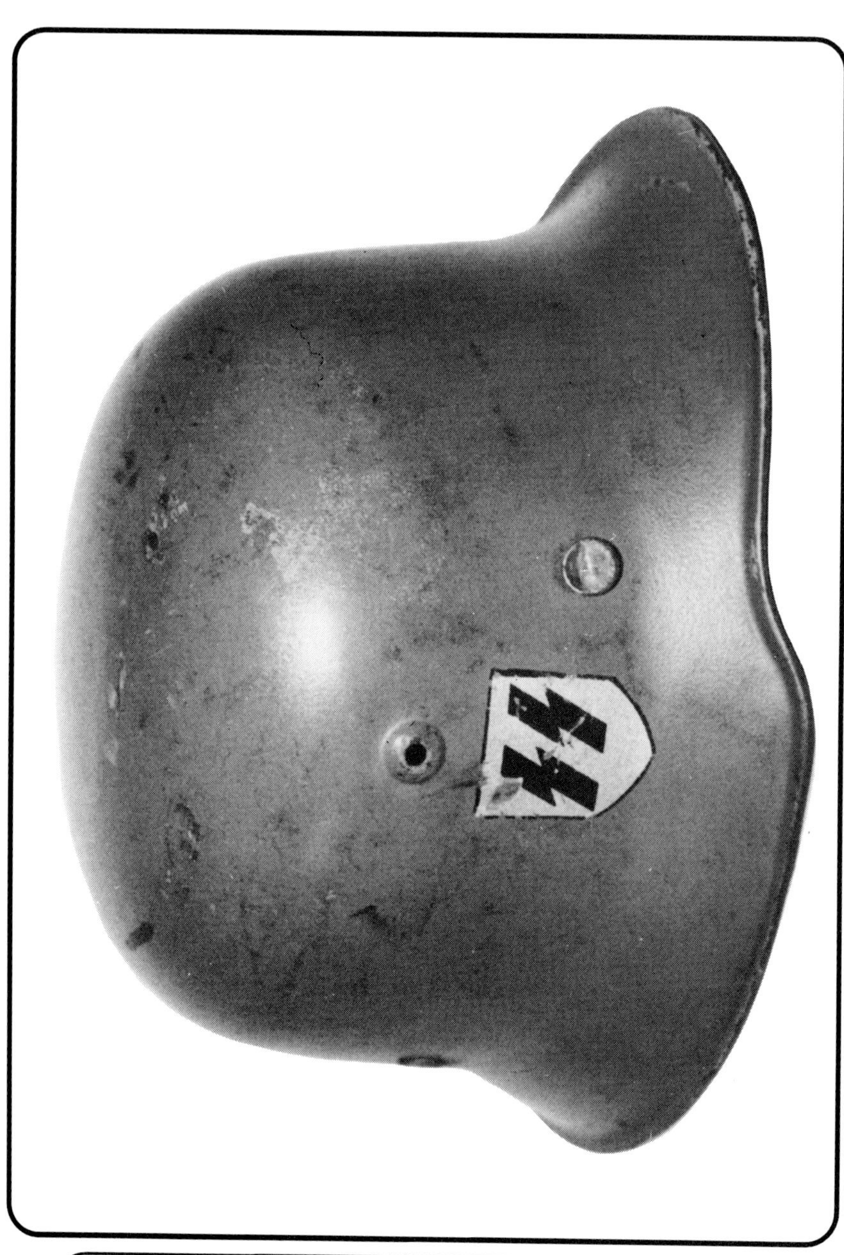

M-34 Medium duty helmet commonly used by the Sicherheitsdienst (SD). This particular helmet has a set of SS decals in reverse or "police" configuration underneath the properly positioned outer decals. Many SD helmets display reversed decals. **Hicks collection**.

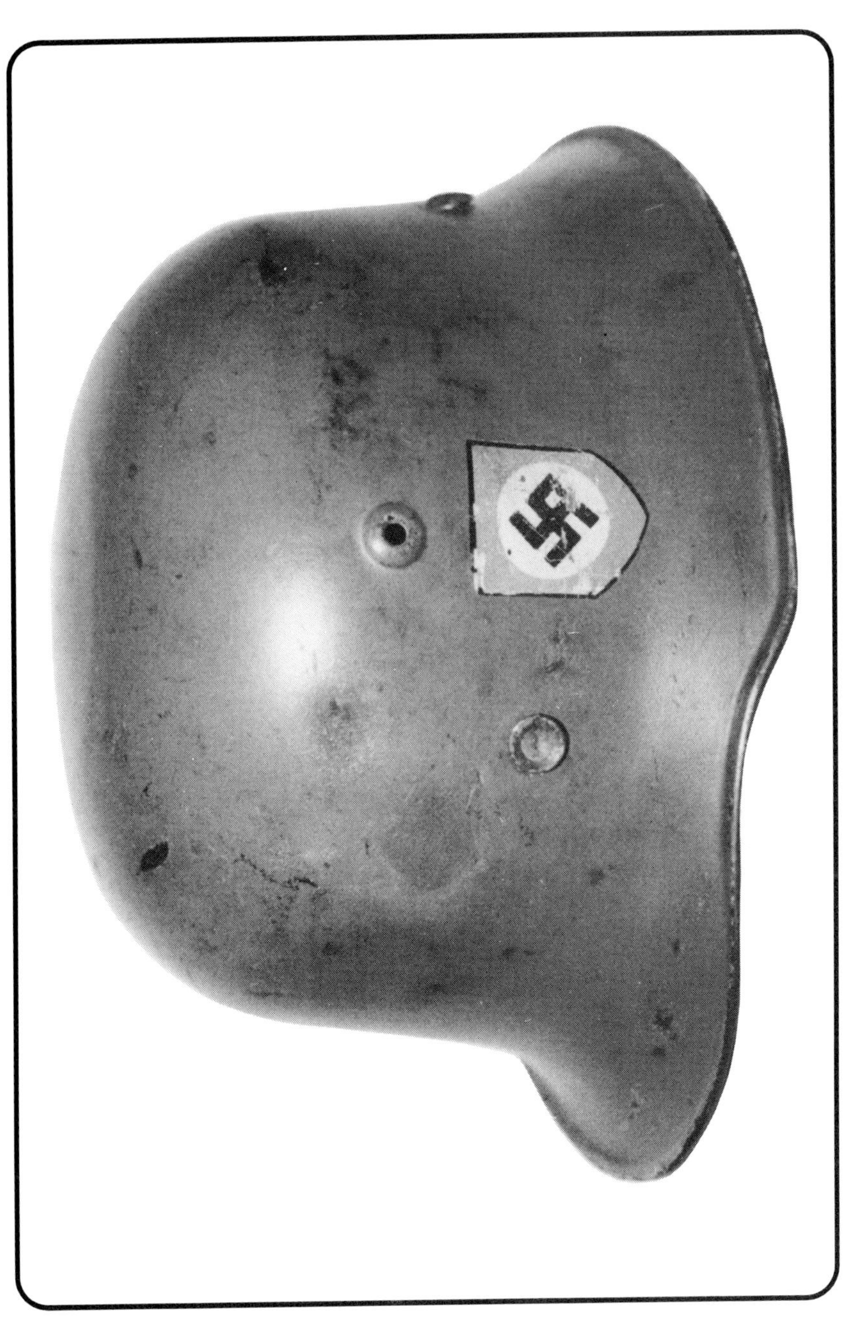

Left side of the M-34 SD medium duty helmet. **Hicks collection.**

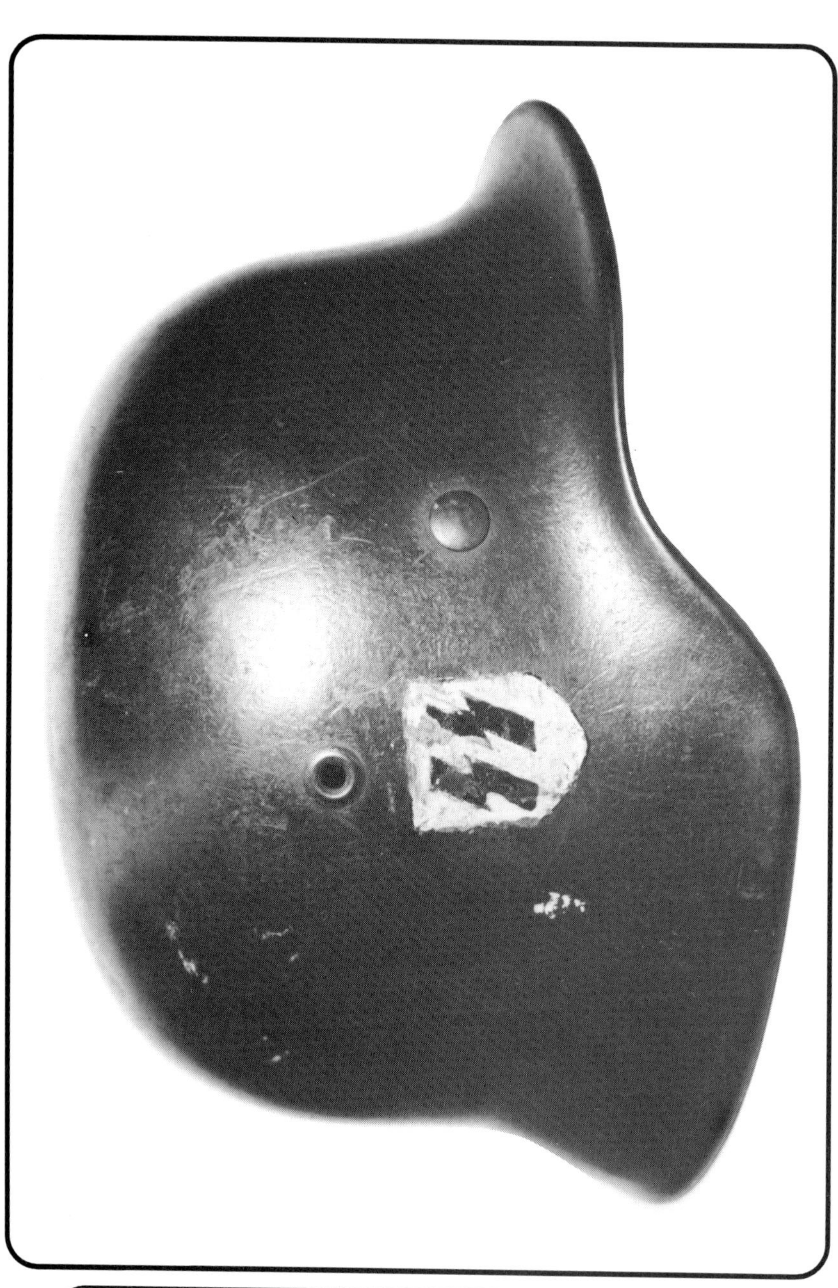

Interesting SS single decal M-40 helmet with radium paint applied over the 1st pattern runic shield, leaving the runes exposed. The helmet was likely used by the SS-Feldgendarmerie. **Al Barrows collection**.

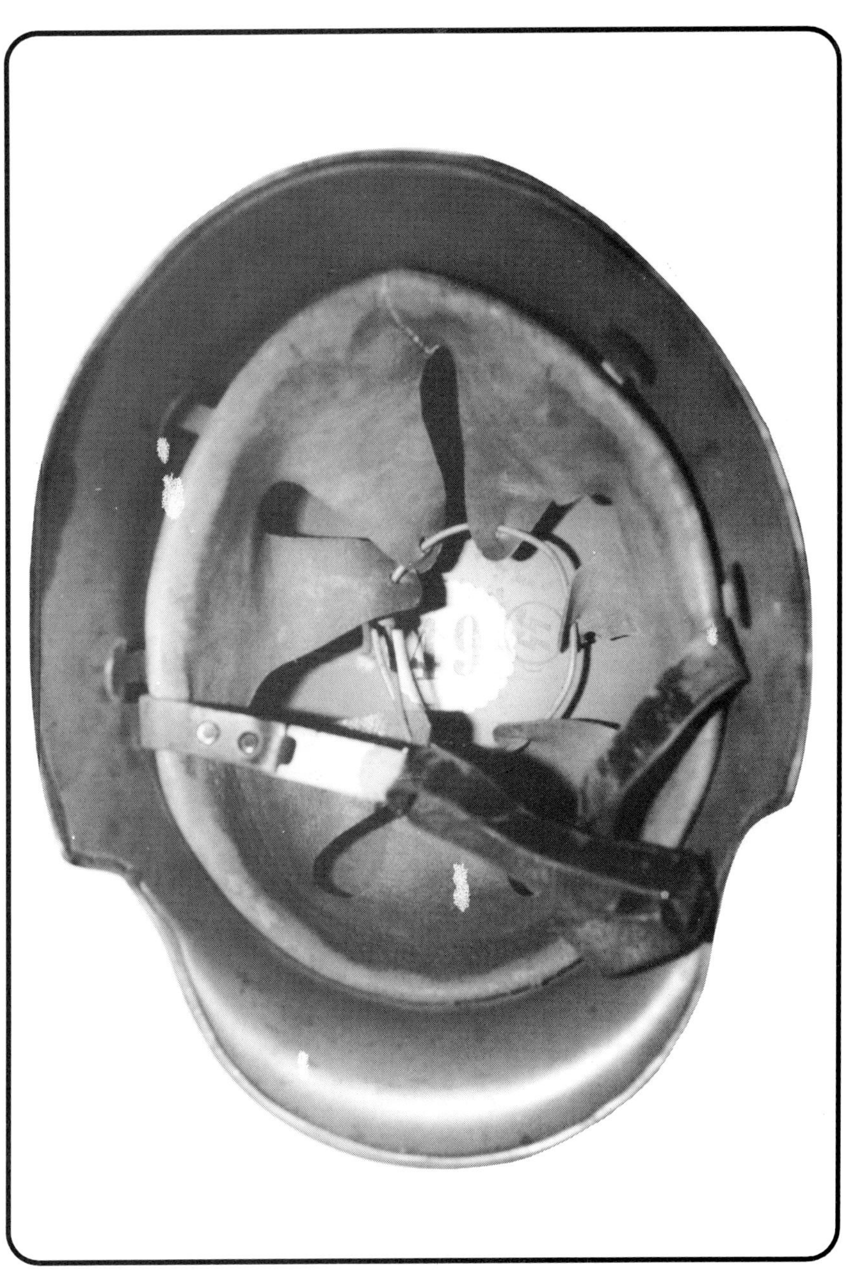

Interior view of civic style SS helmet, showing unusual markings and "rack" number, indicative of some guard or auxiliary role. **Al Barrows**.

BIBLIOGRAPHY

John R. Angolia, **Cloth Insignia of the SS,** BenderPublishing, San Jose, 1983.

Ludwig Baer, **The History of the German Steel Helmet, 1916-1945,** Bender Publishing, San Jose, 1985.

Roger Bender and Hugh Taylor, **Uniforms, Organization and History of the Waffen-SS, Vol. 1-5,** Bender Publishing, San Jose, 1971.

Brian L. Davis, **Waffen-SS,** New York, 1986.

T.V. Goodapple and R. J. Weinand, **German Helmets, 1933-1945; a Collector's Guide, Vol. 1 & 2,** Quincy, Illinois, 1981.

Andrew Mollo, **A Pictoral History of the SS, 1923-1945,** New York, 1976.

Andrew Mollo, **Uniforms of the SS, Vol. 1-7,** Fourth Edition, London, 1991.

Munin-Verlag GmbH, **Wenn Alle Bruder Schweigen,** Osnabruck, 1975.

Alan Wykes, **Hitler's Bodyguards, SS Leibstandarte, New York, 1974.**

SS HELMET AVAILABILITY GUIDE

TYPE . . . . AVAILABILITY

Allgemeine-SS Transitional with
hand-painted insignia (all types) . . . . . Extremely rare

Allgemeine-SS Transitional
(standarddecals) . . . . Very rare

SS-VT Transitional (field grey)
with standard decals . . . . Very rare

SS-TV Transitional (field grey or earth-
brown), variant or "mirror" decals   . . . Only 2 known

Allgemeine-SS M-35 helmet . . . . Very rare

Allgemeine or Waffen-SS "thick runes"   . Only 2 known

Waffen-SS M-35 field grey (1st or 2nd
pattern runes) . . . . Rare

Waffen-SS M-40 double decal helmet . . Extremely rare

Waffen-SS M-40 single decal helmet . . . Rare

Waffen-SS M-42 double decal helmet . . Extremely rare

Waffen-SS M-42 single decal helmet . . . Relatively common

SS "former police"; M-35, 40, 42 (with
runic shield over police shield) . . . . . . Extremely rare to
                  Unavailable

Double runic SS helmets . . . . Less than 10 known

SS-VT / SD M-34 medium duty helmets . Rare

SS Paratrooper helmets . . . . Undeterminable

ABOUT THE AUTHOR

Kelly Hicks is a Major in the US Army Special Forces. He has served in Asia most of his career, and holds a Master of Arts in East Asian Studies from Harvard University. He has been collecting German Helmets since 1964.